FoOTPRINTS IN STONE

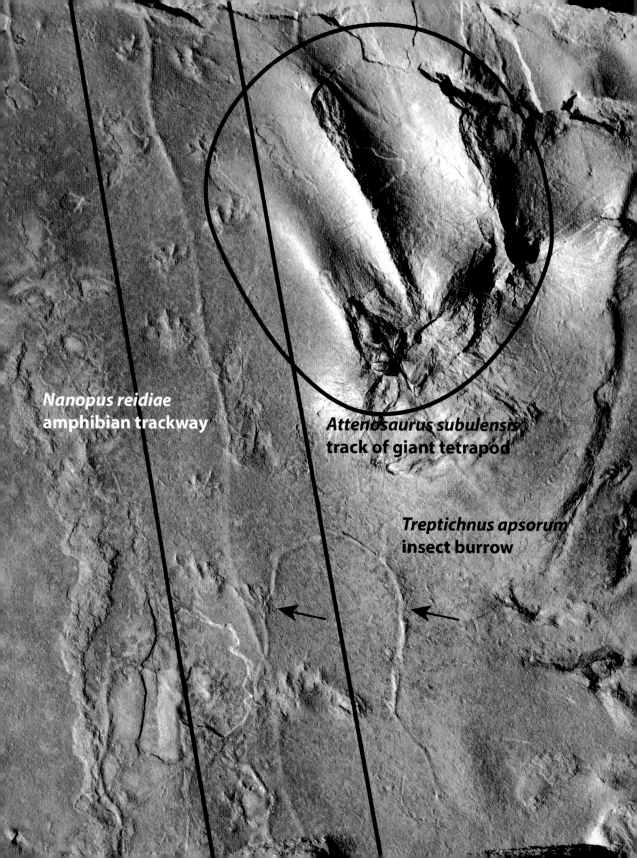

Nanopus reidiae
amphibian trackway

Attenosaurus subulensis
track of giant tetrapod

Treptichnus apsorum
insect burrow

RONALD J. BUTA & DAVID C. KOPASKA-MERKEL

FOOTPRINTS IN STONE

Fossil Traces of
Coal-Age Tetrapods

The University of Alabama Press   Tuscaloosa

The University of Alabama Press
Tuscaloosa, Alabama 35487-0380
uapress.ua.edu

Typeface: Garamond Premiere Pro

Manufactured in Korea

Cover photograph: Amphibian trackway (*Nanopus reidiae*) next to
large track of a reptile-like amphibian (*Attenosaurus subulensis*) with
a small insect burrow (*Treptichnus apsorum*) overlapped by an older
Nanopus trackway; photo by Ron Buta, courtesy of the Alabama
Museum of Natural History.

Frontispiece: Amphibian trackway (*Nanopus reidiae*, in box) next to
large track of a reptile-like amphibian (*Attenosaurus subulensis*, cir-
cled). Note small insect burrow (*Treptichnus apsorum*, two arrows). At
the left arrow, a small *Nanopus* forefoot track overlaps the burrow. The
Attenosaur track is probably younger than the small trackway because
the large track is a moderately deep undertrack whereas the *Nanopus* is
close to a surface track.

Cover and interior design: Michele Myatt Quinn

∞

ALABAMA
Museum of Natural History
THE UNIVERSITY OF ALABAMA

Publication supported in part by funding from the Alabama Museum
of Natural History, the Paleo Alabama Project (Jim and Faye Lacefield), the Alabama Paleontological Society (Prescott Atkinson, MD),
Deborah Crocker, PhD, Bruce and Mary Ann Minkin, the McWane
Science Center, and the Alabama Geological Society.

Cataloging-in-Publication data is available from the Library of Congress.
ISBN: 978-0-8173-5844-0
E-ISBN: 978-0-8173-8956-7

CONTENTS

FOREWORD

The fossil record in the state of Alabama is very rich and much more varied than most amateur and many professional paleontologists realize. With over five hundred million years of geologic time exposed in Alabama, the wealth of material, both body and trace fossils, is almost overwhelming to me as the curator of paleontology for the Alabama Museum of Natural History. Of all the fossils that are found in the state, the Pennsylvanian-aged trace fossils of the Black Warrior Basin in Walker County, Alabama, remain one of the most intriguing and plentiful sources of interest and research for amateur and professional paleontologists alike.

As early as the 1920s, both local and national researchers recognized the scientific importance of the trace fossils found within the Mary Lee coal zone of the Pottsville Formation. American Museum of Natural History paleontologist George Gaylord Simpson even worked to secure specimens for his institution back in New York City. Truman H. Aldrich and Walter B. Jones published the first formal description of these Pennsylvanian trace fossils in 1930 through the Alabama Museum of Natural History Museum Papers (Aldrich and Jones 1930). While specimens have been continually uncovered since that time, primarily by commercial mining of the Mary Lee coal zone, they have received little interest or research until much more recently.

Beginning in 2000, members of the Alabama Paleontological Society, working with the State Lands Division of the Alabama Department of Conservation and Natural Resources, and with Dolores Reid, owner of the New Acton Coal Mining Company, actively campaigned to preserve a Pennsylvanian trackway site in Walker County, Alabama, known locally as the Union Chapel Mine. This site has been referred to as a *Lagerstätte*, or a locality that exhibits exceptional preservation in quality, quantity, and diversity of fossils. Officially designated a facility of the State Lands Division in 2005, the Steven C. Minkin Paleozoic Footprint Site (the Union Chapel Mine) opened a new chapter in the study and collection of Pennsylvanian trace fossils in Alabama. The Alabama Paleontological Society invited researchers from all over the world to participate in collecting and describing fossils found at the site. The result was a monograph entitled *Pennsylvanian Footprints in the Black Warrior Basin of Alabama* (Buta, Rindsberg, and Kopaska-Merkel 2005), which showcases the vertebrate and invertebrate trackways, insect wing and spider impressions, and fossil plants from the locality.

But the story does not end there! I believe that we are entering a "golden age" of Alabama ichnology (the formal study of trace fossils). The Alabama Paleontological Society is still collecting at the Steven C. Minkin Paleozoic Footprint Site, mining activities within the state keep uncovering new deposits, amateur paleontologists keep turning up new and exciting specimens, and the Alabama Museum of Natural History continues to grow our collections. Dr. Ron Buta, coauthor of this book, discusses one such discovery that he made himself. The Crescent Valley Mine locality is a new trace-fossil site that Dr. Buta found in Walker County, Alabama, that is a perfect example of how any person with an interest in fossils and knowledge of the local geology can make a profound contribution to science. So far, he has collected and donated close to two thousand specimens to the Alabama Museum of Natural History.

Footprints in Stone: Fossil Traces of Coal-Age Tetrapods is a wonderful introduction for both amateur and professional paleontologists interested in learning more about the Pennsylvanian-aged fossils from Alabama. The authors take readers through the identification and terminology of trace fossils, life during the Carboniferous, the state of the Earth and the Milky Way at that time, and a showcase of the amazing fossils found in the state. In addition, the authors also explain the discovery and the steps taken toward the eventual preservation of an important fossil locality (the Steven C. Minkin Paleozoic Footprint Site). The information in this book is a valuable resource for both amateurs and professionals who are interested in protecting these nonrenewable resources. Ron Buta and David Kopaska-Merkel have contributed significantly to the study of ichnology with their book.

DANA J. EHRET
Curator of Paleontology
Alabama Museum of Natural History

PREFACE

The past is a never-ending source of fascination for people. Whether the recent or the distant past, we are curious about what the world was like before we were born. We can imagine transporting ourselves back in time and seeing firsthand the events we have read about in history. Imagine, for example, going back to 1863 to hear Abraham Lincoln give the Gettysburg Address, or to 2560 BC to witness the construction of the Great Pyramid of Giza. Human history seems boundless. But as humanity has grown and matured, we have come to realize that the history of our planet extends far beyond the history of human experience. Life existed on our planet long before humans came onto the scene. In the walk of life on Earth, we are relative newcomers.

Sometimes the really remote past comes to us in our own backyard. In Walker County, Alabama, we find evidence of animals and plants that lived in that area so long ago that if we were to transport ourselves back to that time, the world would be alien to us. The landscape would harbor a huge mountain range, the night sky would have different stars, the climate would be much warmer, the sun would be a little dimmer, and the moon a little closer. We would see very large insects and tall, unusual trees with greenish trunks covered with diamond-shaped scales. There would be a deep sea where none exists today. And near the shore of that sea, in swampy, low-lying areas where rivers carried freshwater and sediment from the mountains into estuaries, animals would meander around on wet mud during low tide looking for food, leaving trails of footprints in their wake. From space, we would not see any of our familiar continents, only a single huge landmass. It was a world of life that was already long gone by the time dinosaurs walked the Earth.

This lost world is recorded in the sedimentary rocks of north-central

Alabama. The rocks tell us of major geologic events, including continental collision and mountain building, and of the existence of vast tropical rain forests extending across the supercontinent. The swampy forests, which were so extensive worldwide that the oxygen in the air reached a level of 35 percent (it is 21 percent today), laid down the coal reserves that play an important role in Alabama's economy today. Most interesting is how the rocks preserved the footprints of the ancient animals that lived in the area. These footprints have opened a door to new knowledge of what Alabama was like during the coal age, more than three hundred million years ago.

This is a book about Alabama's *footprints in stone*. The dark gray shale found in Walker County coal mines preserves the footprints, crawl marks, jumping impressions, tail drags, burrows, and resting impressions of a host of primitive creatures, including amphibians, reptiles, insects, fish, and other animals. These organisms lived when reptiles were branching off from amphibians; soon, the new scaly creatures would become dominant land animals. The tracks, known as *trace fossils*, show us these early land animals moving around and going about their day-to-day lives. Trace fossils are fascinating records of past life that have only recently become widely appreciated for what they can tell us about the behavior of long-extinct animals. Trace fossils give us information on ancient life that no skeleton could ever provide.

In 1999, a treasure trove of fossil tracks and traces was discovered at an inactive surface coal mine in east Walker County, the former Union Chapel Mine, located in the small (unincorporated) town of Union Chapel northwest of Birmingham. The fossils were initially discovered not by professional paleontologists but by a high school science teacher who was also an avid amateur paleontologist. He led other amateur fossil collectors to the site. This band of enthusiasts recognized early on that the Union Chapel trace fossils could be scientifically significant and needed to be rescued from reclamation of the mine. Their efforts led to a remarkable amateur and professional collaboration that resulted in three signature accomplishments: the thorough online documentation of the fossils, the publication of a scientific book including original research on the traces, and the preservation of the fossil site by the state of Alabama.

Our goal with this book is to tell the story behind the discovery, documentation, and preservation of the Union Chapel Mine and what it means for Alabama's paleontology and our understanding of what life was like during Alabama's coal age. The story has many facets and represents the experience of people from many walks of life. It is a story that is both worth telling and worth knowing about. In addition to describing how the Union Chapel Mine fossil trackway mother lode was discovered, we

- describe the nature of trace fossils and how they form,
- take you on an imaginary trip back to the coal age,
- describe how amateur collectors worked with professionals and ultimately got the fossil site protected for future study,
- describe and illustrate the many kinds of trace fossils found at the site,
- explain what we know about the animals that made the tracks and show what they might have looked like, and
- describe the findings from a second mine in Walker County, the Crescent Valley Mine, in order to highlight the recent discovery that Walker County likely hosts a megatracksite, an extensive region with multiple tracksites where all the fossils are about the same age.

The book includes numerous illustrations of the types of tracks, trails, burrows, and other traces that have been found at Union Chapel and describes what the traces collectively tell us about the complex ecosystem that once existed in the area.

The book also brings attention to the people who were involved in the discovery and documentation of the first tracks found in Alabama, an event beautifully described by Truman H. Aldrich and Walter B. Jones in a 1930 Alabama Museum of Natural History publication. The tracks had been found in the ceilings or collapsed rocks of tunnels in an underground mine, the No. 11 Mine of the Galloway Coal Company, near Carbon Hill. Aldrich and Jones provided the main names for the exposed footprints, but they were not the only professionals who thought the tracks were important. The discovery attracted the attention of none other than George Gaylord Simpson, associate curator of paleontology of the American Museum of Natural History in New York and one of the foremost evolutionary biologists of the early twentieth century. He visited Alabama in January 1930 to see the tracks firsthand and collect specimens to take back for display and study.

During the seventy years between these first discoveries and those of the amateur collectors, fossil trackway research in Alabama had few adherents in the professional geological and paleontological community. Interest in Alabama trackways increased after the discovery of Permian tracks in New Mexico in the late 1980s by lone "curiosity seeker" Jerry MacDonald. MacDonald became one of the world's best-known citizen scientists for his relentless pursuit of the source of tracks he found in homes and other venues in and around Las Cruces. The success of the venture we describe here owes a great deal to the support and example of MacDonald, who in 1994 wrote a very fine book about his experiences titled *Earth's First Steps: Tracking Life before the Dinosaurs*. Our stories are different but the result is the same: nonprofessional fossil enthusiasts awaken the professional community to the importance of remarkable new trace-fossil discoveries.

A note about measurements: all scientists use the metric system. It is convenient and practical. However, in the United States, nonscientists use the English system of measurement. American scientists use the English system in casual conversation. Therefore, some of the quotes in this book refer to feet and inches. All the measurements we report directly are metric. Not only is the metric system what scientists use, but in some cases we talk about very small things that are more easily measured in millimeters than in inches.

ACKNOWLEDGMENTS

The authors are grateful to Elizabeth Motherwell, natural history editor of the University of Alabama Press, for inviting the Alabama Paleontological Society (APS) to write its story and for her unwavering support during the ups and downs of the project. We thank Ashley Allen, Dr. T. Prescott Atkinson, Gorden L. Bell, Prof. David L. Dilcher, Jun Ebersole, Jim Griggs, Prof. Hartmut Haubold, Dr. James A. Lacefield, Dr. James P. Lamb, Dr. Spencer G. Lucas, Jerry MacDonald, Prof. Tony Martin, Dr. Bruce Minkin, Missy Minkin, Steven C. Minkin Jr., Dr. Nic Minter, Dr. Jack C. Pashin, Dolores Reid, Dr. Andrew K. Rindsberg, John Southard, Kathy Twieg, and Donny Williams for agreeing to be interviewed so that we could tell this story as accurately as possible. We are especially grateful to Dr. Jennifer A. Clack, professor and curator of vertebrate paleontology at the Museum of Zoology, Cambridge University, for her communications and for allowing us to use several illustrations from the second edition of her excellent book *Gaining Ground: the Origin and Evolution of Tetrapods*. We are also grateful to Dr. Jack C. Pashin, professor and Devon Chair of Basin Research, Boone Pickens School of Geology, Oklahoma State University; and Dr. James A. Lacefield, author of *Lost Worlds in Alabama Rocks: A Guide to the State's Ancient Life and Landscape*, for the use of several excellent illustrations.

The book was locally reviewed, in whole or in part, by Ashley Allen, T. Prescott Atkinson, Andrew K. Rindsberg, Deborah A. Crocker, Jack C. Pashin, and Greg and Jenny Smith. We also thank Martin Lockley, Douglas Haywick, and two other anonymous external reviewers for helpful comments and constructive suggestions that considerably improved the manuscript.

We are deeply grateful to Leland Lowery, foreman of the Crescent

Valley Mine (CVM), located near Carbon Hill, for allowing coauthor Buta to extensively explore the mine for fossil trackways and build up a sizable database of specimens that could be directly compared to the Union Chapel database. This was an unexpected delight and a rare privilege that happened because of our interest in having a historical component in this book, and it introduced us not only to active surface-mining techniques but also to active miners. CVM fossils provided a much-needed perspective on the diversity of the Union Chapel fossil assemblage, and also a large volume of new material for future professional paleontological research. We are grateful to miners Tim Richie, Richard Beards, and Donny Williams for contributing several important specimens to the new database.

When the CVM was first opened in 2008, the mining operation exposed old cavities that had been cleared of coal decades ago. In the ceilings of some of these cavities, trackways were seen just as they had been more than seventy years earlier. We are grateful to Donny Williams, foreman of the Kansas Mine No. 2 near Eldridge, Alabama; William Sharpton, geologist of the National Coal Company (owners of the CVM at the time); and Anthony J. "Tony" Edwards, geologist and property manager, vice president, Regions Natural Resource Department, Birmingham, for bringing our attention to these observations. We also thank local geologists William Lawrence, Mark Chapman, Whitney Telle, and Bob Carr for other helpful information on track collecting and mining in Alabama.

We are grateful to the late Lewis Dean, librarian of the Geological Survey of Alabama in Tuscaloosa; Ruth O'Leary, director of Collections and Archives, Division of Paleontology, American Museum of Natural History; and Tamara Braithwaite, registrar of collections, Memphis Pink Palace Museum, for providing documents and information that allowed us to gain a better understanding of George Gaylord Simpson's visit and what he did while in Alabama. We also thank Charles M. Whitson, Alabama Department of Labor, Abandoned Mine Land Division, for communicating his research on the history of Galloway Company mining in Alabama.

Coauthor Ron Buta was delighted, during the course of this work, to serendipitously meet Clarence Blair, former president of Black Diamond Coal Company, at a meeting of the New Horizons learning group in Birmingham. Blair is the nephew of Arthur J. Blair, who in 1929 was the first professional geologist to go into the No. 11 Mine of the Galloway Coal Company to see the tracks that had been reported to adorn the ceilings of some of the mine tunnels.

Ron Buta also thanks John L. Southard, a local resident of Union Chapel, for describing the human history of the area of the Union Chapel Mine, and for taking him on a grand tour of local landmarks. A man of gentle southern charm and natural curiosity, Southard has lived his entire life in Union Chapel, working on the same land his parents, grandparents, and great-grandparents worked, and attending the same church that they attended. The APS, and the Union Chapel Mine project, were extraordinarily lucky to have someone like Southard nearby. Far from being disturbed by the unusual influx of visitors, Southard was fascinated by the fossils people were finding in the area and often went out of his way to meet people and discuss what he knew about the area. His demeanor and easy-going nature made visitors feel very comfortable and welcome in his presence.

The fossil traces illustrated in this book were collected by a variety of people. We gratefully acknowledge these collectors of the following Union Chapel Mine (UCM) numbered specimens: Ashley Allen (UCM 285, 312, 331, 357, 368, 417, 1034), Bruce Relihan (UCM 469, 485k, 953, 1728, 1731), Carl Sloan (UCM 5000b), Gerald Badger (UCM 3780, 3781), Dr. Jim Lacefield (UCM 124), Jay Tucker (UCM 249, 254, 263), Ken Hoyle (UCM 117, 123), Larry Hensley (UCM 1881), Mike Robitaille (UCM 493), Dr. Ron Buta (666, 677, 680, 743, 1153, 1267, 1268, 1272, 1348, 1349, 1368, 1377, 1410, 1488, 1505, 1548, 1735, 1748, 1749, 1758, 1797, 2387, 2449, 2466, 2539, and unnumbered specimen shown in fig. 16.8C), Steve Minkin (18, 24, 26, 206), Dr. T. Prescott Atkinson (109, 437, 451, 637, 1060, 1070, 1071, 1074, 1075, 1076, 1141, 2281, 2369, 2742), and Drs. Andrew K. Rindsberg and David C. Kopaska-Merkel (UCM 1311, 2026, 2038, for the Geological Survey of Alabama collection). Cindy Wallace of Empire, Alabama, found the beautiful specimen shown in figure 15.9D. We are also grateful to Dr. Spencer G. Lucas (New Mexico Museum of Natural History and Science) and Matthew Stimson (St. Mary's University) for allowing us to use their photographs of some of the above specimens.

Finally, we are grateful to the Alabama Paleontological Society, particularly Ashley Allen and Dr. T. Prescott Atkinson, for their support and encouragement during the years it took to produce this book, and to local Tuscaloosa artist Sue Blackshear for her beautiful depictions of the living animals of the Union Chapel Mine, now formally known as the Steven C. Minkin Paleozoic Footprint Site.

FOOTPRINTS IN STONE

1

The Tracks Above

"Rush and I don't mean maybe!!!" wrote Walter B. Jones in the margin of his new manuscript. It was early in 1930, just a few months after he and his staff at the Alabama Museum of Natural History had participated in what would be one of the most important discoveries during his tenure as Alabama's state geologist: fossil animal tracks in an underground coal mine near Carbon Hill. Workers at the Galloway Coal Company No. 11 Mine, in operation since 1912 and located just a short distance south of Carbon Hill, had seen the curious fossils in various parts of the mine and had informed their managers about them. The large underground slope mine had yielded tracks for several years, particularly in an area called the "Southwest Slope," located more than a mile from the mine entrance, but a particular exposure appeared to draw more attention than previous finds: a forty-foot-long trackway of a mysterious and relatively small five-toed animal with strange outward-projecting outer hind toes adorning the *ceiling* of one of the mine tunnels. Mine general manager W. Frank Cobb Sr. and chief engineer A. P. McIntosh were informed of the tracks, and Cobb visited Jones's Tuscaloosa office in late November 1929 to talk about them. The tracks appeared to come from rock layers lying within a meter of the top of the coal bed being mined. They'd been exposed because the shale containing the tracks would often collapse to the mine floor or be removed to give greater clearance. Cobb later told Jones that McIntosh could recover a three-by-two-foot slab of tracks from the collapsed rock, and that it would be simple to make a plaster cast of some of the ceiling tracks.[1]

Several months earlier, McIntosh had contacted a local geologist,

Arthur J. Blair, from the Tennessee Coal, Iron, and Railroad Company in Birmingham, about the tracks. The TCI, as the company was called, worked mainly in steel production but had interests in coal and iron ore mining as well. Initially, Blair was skeptical about the tracks, but in early December 1929, he decided to go take a look at the exceptional display. Blair and another company man, Ivan W. Miller, visited the mine and were stunned when they saw that the long ceiling trackway was real. Blair was able to extract part of the trackway and took it to his office in Birmingham, where the young and relatively newly appointed Jones was waiting when he brought the slab in. On seeing the incredible specimen, Jones declared the tracks to be "entirely new to Alabama geology."[2]

The find generated considerable interest among staff at the Alabama Museum of Natural History and the University of Alabama, where Jones had his office. Both Jones and assistant museum curator David DeJarnette, as well as other museum staff, made special visits to the mine to collect more tracks. Cobb wrote to Jones telling him that the editor of the journal *Coal Age* wanted him to write a short article on the tracks, pointing out that the magazine had published articles about the Galloway mine before.[3] Jones agreed to write the article quickly and informed Cobb that he wanted to send a group up to the mine soon for a few more days, including himself and the museum's curator of paleontology, Truman H. Aldrich Sr.[4] Jones also told Cobb that he wanted to issue an account of the tracks as a museum paper in order to keep credit for the discovery entirely in-state; hence the need to rush publication. To aid in writing the paper, Jones received information from some outside sources, including the famous American Museum of Natural History paleontologist George Gaylord Simpson, who provided a bibliography of fossil footprints from the same time period. The resulting study was submitted in April 1930 and published later that year as *Footprints from the Coal Measures of Alabama*, Museum Paper 9, under the authorship of Aldrich and Jones. Jones wrote the introduction and background, while Aldrich named the tracks and provided their scientific description.

The tracks Aldrich and Jones described were made by tetrapods: vertebrate (backboned) animals with four limbs. Early tetrapods included only amphibians and reptiles, but today the class also includes mammals and birds, as well as any animal descended from four-limbed ancestors (such as snakes). The animals lived more than three hundred million years ago during what is often referred to as the "coal age," or Carboniferous Period. This was a time when the world's tropical forests grew explosively, producing the plant material that created the coal reserves now being mined in Alabama and elsewhere. Jones believed the Alabama tracks were made by some of the oldest known land vertebrates, including both amphibians and reptiles, and noted that while tracks of amphibians had been found in rocks much older than those in the Alabama mine, the same was not true of reptilian tracks. The find was clearly exciting to him for this reason, and although he could not take full credit for the discovery (he gave that to Blair and Miller, not for finding the first tracks, but for being the first to seriously look into them), he could take credit for documenting them and bringing them to the attention of a wider number of specialists.

Although the No. 11 Mine yielded abundant footprints of the ancient coal-age tetrapods, not a single bone or skeleton was found, making it difficult to know what the animals looked like. Jones concluded in Museum Paper 9 that as a result, "the tracks of these ancient animals have been of little benefit to the anatomists and morphologists in the United States." In a press release dated January 24, 1930, George Gaylord Simpson, who had visited the No. 11 Mine to collect specimens himself, stated that the "discovery is not sensational, but it is of great scientific importance and popular interest and is worthy of careful study and of preservation."[5]

The idea that the discovery of the Alabama tracks was important but not "sensational" has to be viewed in the context of general paleontological discoveries and of the time when they were found. The tracks would have been more sensational if they had been the *oldest* reptilian tracks ever found, or if they had been the *first* found from that time period. But other sites had been previously identified that matched the age of the Carbon Hill specimens, such as the Fossil Cliffs of Joggins, Nova Scotia, which had been known since the mid-nineteenth century. Perhaps in part because of this, and because the study of fossil footprints was still a relatively new field, Museum Paper 9 was not followed up by any further studies and was soon forgotten by most, although the

FIGURE 1.1 Tracks of a large reptile-like amphibian collected by Aldrich and Jones from the No. 11 Mine after their main paper had been published. The specimen is a trackway called *Attenosaurus subulensis*, which shows hindfoot tracks only and was on display at the Alabama Museum of Natural History until 2004.

Alabama Museum of Natural History displayed some of the tracks until as recently as 2004 (fig. 1.1).

Fast-forward now seven decades, from 1929 to 1999. By this time, the Birmingham Paleontological Society (BPS) had been in existence for fifteen years. The amateur fossil-collecting group, founded by former staff members of the Red Mountain Museum in Birmingham,[6] regularly visited coal mines to collect coal-age plant fossils, which are some of the most beautiful fossils found in Alabama. The carbonized remains of extinct seed ferns, which have no modern counterpart, and the bark impressions of peculiar scaly trees, also unmatched in modern times, are a huge draw to the mines. Sometimes whole plants with multiple fern fronds and three-dimensional casts of tree trunks can be found. Coal-age rocks preserve a record of plant life and environmental conditions almost unmatched by any other period in Earth's history.

The BPS had regular monthly meetings in the Homewood Library on Oxmoor Road. In addition to hearing a speaker, members often described or showed the fossils they had collected on the most recent BPS field trip. None of these finds sparked the group to take the initiative to study the material, because the kinds of fossils found had generally been well studied by professionals many years before. At the December 1999 meeting, however, a BPS member presented to the group fossils he had collected at a new site, a surface coal mine just east of Jasper, Alabama. The fossils were coal-age tetrapod footprints, something many of the members had never seen before. The footprints appeared to scurry across dark gray slabs of shale, and you could tell which way the animal was going. The tracks were also uncharacteristically small. What was most intriguing to those who saw the tracks was that they were not the remains of a once-living entity, like bones, but were instead natural recordings of an animal going about its daily activities when it was alive. That is, the tracks were about the life of an animal, not its death.

Not surprisingly, most of the amateurs who saw these specimens were unaware of the earlier Carbon Hill discoveries, but once this group saw the tracks, they were bound to be taken seriously. Suddenly, the forgotten Alabama tracks were noticed again. This sparked a chain of events that would return the ancient footprints of Walker County to the forefront of Alabama paleontology and change the amateur fossil-collecting group in a big way. Who was the collector who brought in these tracks, and where did he find them?

DISCOVERY IN UNION CHAPEL

Ashley Allen could not believe what he had just heard. It was an ordinary November day at Oneonta High School in Blount County, Alabama, and he was talking to his seventh-grade class about plant fossils he had collected from surface coal mines just west and southwest of Birmingham. The twenty-nine-year-old (and later award-winning) science teacher, a 1992 graduate of the University of West Alabama (UWA) and also an active member of the BPS, enjoyed collecting fossils, and in the fall of 1999 he was looking for a new mine suitable for a class field trip. He wanted his students to find and identify fossils on their own and learn about ancient life through direct experience, instead of just looking at what someone else had found. Then, unexpectedly, student Jessie Burton raised his hand and said his grandmother owned a coal mine! Stunned, Allen at first thought Jessie was joking, replying "Yeah, okay, right Jessie. Let's go on," but young Jessie insisted it was true. Dolores Reid, Jessie's grandmother, recalled that her grandson had a hard time convincing Allen that she owned mines (D. Reid, pers. comm.).

Now, the story could have ended here. Owners of coal mines are not always amenable to letting people in to look for fossils. Surface mines are generally treacherous places because of the rugged terrain (usually dominated by huge piles of rock), and liability for injury is a serious issue to consider before letting anyone into a mine, much less a class of seventh graders. Some mine managers and owners would simply not allow it. But when Allen phoned Reid, she said she would love for him to come out and scout one of her mines. Not only did she see the educational value of such a field trip for the kids, but her grandson was thrilled about being able to go, and she wanted to make him happy.

Reid phoned her superintendent and set the wheels in motion for Allen to visit one of her mines, followed by a field trip for his class. Under ordinary circumstances, Allen and his class would have found mostly plant fossils. After all, coal is nothing more than highly compressed plant material that has turned black and been amalgamated under pressure and heat. But Allen's experience as an active member of the BPS, and as a former geology student at the UWA, gave him knowledge of coal-mine fossils that went beyond mere plants. He knew that coal mines could yield fossil tetrapod footprints, particularly the footprints of primitive amphibians, which would have been abundant at the time the plants were alive. Allen did not expect to find any, but in the back of his mind he knew such things could turn up, and he was fascinated by the prospect. His UWA geology teacher, the late Prof. Richard Thurn, had told him to keep a lookout for tracks, because "a few had been discovered decades ago." Since he was about to visit a mine that amateur collectors had not visited before, Allen was excited about what he might find. Could there be footprints?

The Union Chapel Mine (UCM) was in active operation by Reid's company, New Acton Coal, in late 1999. The locale, Union Chapel,[7] is a relatively small grouping of houses and a church (an unincorporated community) just southeast of Jasper, Alabama, off Highway 78.[8] The mine superintendent had told Reid that they saw more fossils at the UCM than at any of her other mining properties. Allen thus arranged a visit to the UCM in order to see what was there and make sure it was safe enough for a class field trip.

The scouting trip took place in November 1999 on a cool and sunny day, ideal, as it turns out, for seeing trackways. When Allen arrived at the site, he met the mine foreman, who told him where he could safely look for fossils. Basically, he was directed to stay away from the areas where the miners were actively working with heavy equipment, and from any blasting area. Perhaps to whet Allen's appetite, the foreman showed him a fossil tree trunk. This was a natural cast formed by the rapid burial and decay of the tree, and subsequent in-filling of the decayed trunk's cavity by mud or sand that later solidified. In life the tree was a huge lycopod with a beautiful pattern of elevated cushions called leaf scars. Such trunks are not rare at coal mines but are often too heavy to carry out or too dangerous to retrieve because they are still in position in a vertical cliff ("highwall") created by the mining operation.

After seeing the trunk, Allen was given free access to the rock piles and he started to look around on his own. Initially, he stayed near the mine road and looked at the lower parts of the rock piles. After only a few minutes, he found a coffee-table-sized slab with a definite trackway (fig. 1.2, top right). From the texture and color of the rock, it appeared that the animal that made the trackway had walked in soft, very fine-grained gray mud. The trackway was well preserved and Allen thought it may have been made under a thin film of water. As interesting as it must have been to see it, the trackway did not look as if it had been made by

FIGURE 1.2 *Top left*, Ashley Allen talks about horseshoe crabs to his science class at Oneonta High School, 2012; *top right*, the first tracks discovered at the Union Chapel Mine, most likely from an invertebrate animal; *middle*, the second trackway discovered at the Union Chapel Mine, likely from a horseshoe crab; *bottom*, the first tetrapod tracks discovered at the Union Chapel Mine, likely from a small amphibian. The arrows show the directions of three different sets of tracks. Scales in cm for all photos in this book except as noted.

a tetrapod. If it had been, he would have seen footprints with toes and possibly a tail-drag mark. Instead, it looked like "tire tracks of a small truck or little radio-operated vehicle" or, more scientifically, the tracks of an invertebrate (nonbackboned) animal. All he could really say about it at the time was that it was a *trackway of something*.

Allen had been looking for fossil trackways in coal mines for several years prior to his excursion to the UCM, having visited about a dozen other sites with little success. At the (now reclaimed) large Cedrum Mine (near Cedrum, Alabama), for example, he found only one small trackway. He also saw a small trackway when he visited a mining engineer in Tuscaloosa around this same time. Those were the only tracks from Alabama he had seen prior to his UCM visit. Finding a trackway of something meant that *a mechanism of forming and preserving tracks* existed at the UCM. If tetrapods actually did live in the area, they would have left tracks also.

Continuing with his scouting, Allen found another trackway (fig. 1.2, middle), a type he later thought had been made by a primitive horseshoe crab rather than a tetrapod. Horseshoe crabs are not really crabs but are more closely related to spiders and scorpions. They leave distinctive and complex trackways owing to their five pairs of walking legs (chapter 16). Although still not a tetrapod trackway, this second specimen again pointed to a mechanism for preserving tracks at Union Chapel, and Allen was getting more excited about what else he might find. He walked a short distance to another area by a road that looked "pushed through the rocks" and spotted a long, thin slab about the size of two books stacked end to end. He decided to split the slab. Many of the slabs lying around were thinly layered and would come apart like the pages of a book, either with a little tapping with a hammer and chisel or just by pulling the layers apart. When he split this particular slab open (fig. 1.2, bottom), he was stunned to see *three beautiful tetrapod trackways*, which at the time he thought were made by primitive amphibians. There was no doubt—these tracks were made by a tetrapod. The tracks were well-formed footprints with multiple toes that ran in three different directions on the slab. Seeing the tetrapod tracks, he was so excited that he yelled "Yahoo!" almost loud enough for the miners to hear on the other side of the mine (A. Allen, pers. comm.). This was the moment he had been waiting for!

Before he went home that afternoon, after perhaps as much as six hours of searching, Allen had discovered a dozen tetrapod trackways, as well as plant fossils and trackways made by invertebrates. All the fossils were beautifully preserved. One specimen was so nice that he ran up to a big truck with several miners and said, "Hey look, it's amphibian trackways!" They looked at him like he was a madman and said, "Yeah, it's fossils!" From their reaction, Allen thought the miners were simply used to seeing fossils at the mine and as a result were not too impressed with what he showed them. Still, Allen could not contain his excitement. He was on cloud nine from such a productive day's effort.

The fact that the first brief visit yielded so much suggested a rich site for further exploration. Allen had clearly tapped only the surface of what might be out at that site. Little did he know that these humble findings would lead to the UCM becoming the first state-protected fossil site in Alabama. The timeline of events following Allen's discovery is summarized in appendix 1.

After his successful initial visit, Allen took his class to the mine. Jessie's grandmother accompanied them. Reid had gotten into the mining business through her late husband, who was involved mostly in underground mining. The land for UCM had been leased to her for surface mining, but the owners really wanted to sell it and she eventually bought it. Although she owned more than one coal mine and had seen whole fossil trees and other plant impressions, this did not mean she was an expert on the fossils frequently found at such mines. On the day she went out with Allen and his class, she was amazed at how much there was that she had never noticed before. She started to see fossils everywhere. On that same day, a big chunk of the highwall fell off all of a sudden and made a loud boom. It did not fall near any of the students or miners, but it was a graphic reminder of how dangerous a surface mine could be. She made sure all the kids stayed close and in a group. To date, no one has been struck by falling debris at the UCM, mostly because the highwall area is now cordoned off.

Verifying the Discovery

Following his initial visit to the mine, Allen contacted Dr. James A. "Jim" Lacefield, author of *Lost Worlds in Alabama Rocks: A Guide to the State's Ancient Life and Landscape*. Jim is a local expert on Alabama fossils. He earned a PhD in science education at the University of Alabama, specializing in paleobiology and paleoecology. He wrote *Lost Worlds* when he realized that there was no comprehensive resource teachers could turn to for information about ancient environments and ancient life in Alabama. He synthesized into a beautiful, illustrated book what geologists knew about ancient Alabama, presenting it in plain English that everyone could understand (J. Lacefield, pers. comm.).

Over the years, Lacefield had collected tetrapod trackways from other coal mines in Alabama. The previous summer he had given the

BPS a lecture about the coal age and had shown photographs of track-ways he had found at Alabama coal mines, including one he had visited less than ten years earlier near the small town called Kansas, only a few miles west of Carbon Hill. He confirmed that Allen had indeed found the same kind of thing at the UCM. Trackways made by both vertebrate and invertebrate animals have been preserved in association with most coal-bearing rocks in Alabama, but two characteristics made the UCM unusual. First was the sheer quantity of well-preserved traces, especially vertebrate trackways. This was apparent almost right away, because so many specimens were found, even that first day. Second was the diversity of the fossils. The site yielded more than one type of tetrapod trackway and also several types of invertebrate traces. The richness of the fossil assemblage pointed to a complex ecosystem that could support a significant variety of animals and plants.

After Allen's scouting visit, it wasn't long before the December 1999 BPS meeting took place, at which he brought two or three specimens and laid them out on a table. After seeing such amazing fossils, society members naturally asked about the possibility of a group field trip to the mine. Allen once again phoned Reid to ask her permission, and she said absolutely, they could go and they should enjoy themselves. With that kind of enthusiastic support, the group visited the site on January 23, 2000.[9] Each visitor had to sign a form absolving the mining company of any liability for injury, something that most mine owners require before allowing entry to a site. Some of the remarkable fossils found during the BPS's first visit are described in chapter 8.

As excited as Allen was about finding fossil tetrapod footprints the day of his scouting visit, he was of course at the site mainly as a teacher and an amateur fossil collector, not a trackway researcher, and was not aware of how significant his discovery would turn out to be. The earlier studies of Alabama tracks by Aldrich and Jones had been largely forgotten,[10] and even though more track discoveries had been made at other coal mines between 1929 and 1999 (as Jim Lacefield had noted), no new knowledge had come out of any of these finds. It was not until Allen's find that the tracks were once again seen by people who knew what the fossils meant and who were aware of the possible new knowledge that could come from studying them in detail. Allen's knowledge, his awareness of the possible presence of tracks in Alabama coal mines, and his communication of the find to his fellow amateur collectors in the BPS were crucial to the discovery getting the attention it deserved. One of these other amateur collectors, Steven C. Minkin, was so convinced of the scientific importance of the fossils that he initiated a documentation and preservation effort that changed the way amateur paleontology was done in Alabama. He insisted that the trackways had to be brought to the attention of professionals, especially those who specialized in the study of *trace fossils*, a scientific field known as ichnology, from the Greek word *ichnos*, meaning "trace." Trackways are trace fossils because they are the traces of activities of ancient animals, made in wet mud or sand and preserved in rock. The UCM, renamed the Steven C. Minkin Paleozoic Footprint Site in 2005 in honor of the late BPS member's contributions (chapter 13), is now recognized as the most important fossil trackway site of its age in the world (chapter 12).

To fully appreciate the importance and significance of Allen's discovery, we need a little more background information. Exactly how old are the UCM trackways, and how has that age been determined? How were they exposed? What is the nature of the rock they were found in? What types of animals made the tracks? We examine these issues in the next several chapters.

2

A Lost World

Footprints are a fact of life on a world with numerous walking creatures and an abundance of wet mud. Any animal that ventures onto wet mud will leave behind a trail of footprints or other impressions that record its activities. Often, a given creature will leave a characteristic set of tracks that allow the trackmaker to be identified. For example, deer have double-toed hooves, and their tracks are quite distinctive. This does not mean that tracking is always unambiguous, because different animals can sometimes leave similar-looking footprints (Murie and Elbroch 2005). Mud is also not the only possible track-recording material. Animals walking in snow or sand can also leave tracks.

Footprints can be preserved as fossils. The most famous footprint fossils are those made by dinosaurs. Hundreds of dinosaur track sites are known worldwide, and many of these sites are well documented (Lockley and Hunt 1995; Lockley and Meyer 2000). The beauty of dinosaur footprints is what they tell us about how dinosaurs behaved: some moved in herds, while others engaged in predatory behavior. But as interesting and popular as dinosaur footprints are, there were creatures walking the Earth long before dinosaurs. These creatures also left footprints in wet mud that were sometimes preserved as fossils. And like dinosaur footprints, the trails of these ancient smaller animals tell us something about how they behaved and interacted with their environment.

Fossil trackways can take us directly into the mysterious and often-times alien world of ancient animals. It is fascinating to see the preserved trail of a creature that walked across a muddy surface hundreds of millions of years ago. Tracks are like a snapshot of life that had no other way of being recorded. A pile of fossil bones can tell us something about what an ancient animal looked like, but very little about how it might have behaved. Tracks complement our knowledge of the prehistoric world implied by skeletons and other body fossils.

The vast majority of footprints and trails of living animals are not preserved at all. Usually, footprints are simply washed away the next time the ground gets wet, or they are eroded away later. Special conditions are needed to preserve them as fossils. If a trackway can be buried quickly and deeply enough, it can be preserved for millions of years. If this happens, it has a chance of being discovered as a trace fossil. The study of trace fossils—evidence of past life that is *not* the remains of the organisms themselves, but preserved traces of their life activities—was still a relatively new field in the 1920s when the first tracks were discovered in Alabama. The significance of trackways for understanding ancient environments and ancient life was not appreciated until many years later.

FOOTPRINTS IN ALABAMA STONE

All the tracks that Aldrich and Jones photographed and named, and most of those that Ashley Allen and members of the BPS found later, were impressed in a fine-grained, dark gray laminar shale, a common type of sedimentary rock that naturally occurs in flat and relatively thin layers. It is ubiquitous in Alabama coal mines and was laid down during the "coal age," the time in Earth's history when dense tropical forests that grew for thousands of years dominated the landscape. Coal represents the carbonized, highly compressed remains of ancient swamp plants. Coal forms beds (seams), centimeters to meters in thickness, which have been compressed by a factor of about ten from once-living plant material (Pashin 2005). The plants were prevented from completely decaying by being buried in sediment saturated with water containing little or no dissolved oxygen. Like us, the bacteria that break down plant and animal material need oxygen to live, but if plant material accumulates in oxygen-poor water, it can be transformed into a dense organic-rich substance known as peat. Over time, high pressure and temperature, caused by the accumulation of heavy layers of overlying sediment referred to as overburden, turn peat into coal.

Coal-age rocks are extremely old. The ones containing trackways in Alabama have been dated at about 313 million years, which is four times as old as the Rocky Mountains.[1] The coal seams of Walker County, which have colorful names such as the Mary Lee, New Castle, Blue Creek, and Jagger, were all once major tropical swamp forests. The forests were episodic, coming and going every hundred thousand years or so. One often sees in the exposed rock layers of Alabama coal mines whole standing tree trunks that were buried in life position, reminders of the power of some ancient storm or other extreme event.[2]

The 313 million years age lies within the broad period of Earth's history known as the Carboniferous, which ranges from 299 to 359 million years ago. The period is so named because much of the world's coal comes from rocks of this age. (Carboniferous means "coal bearing.") To see where the Carboniferous Period fits relative to the rest of Earth's history, geologists make what is known as a "geologic time scale," in which Earth's history is divided up into units representing various periods of time (fig. 2.1A, B). The time scale is displayed as a column ranging from youngest at the top to oldest at the bottom. For example, rocks from the Upper Carboniferous are younger than rocks from the Lower Carboniferous. Geologists in the United States have named the Lower Carboniferous the Mississippian Period, ranging from 318 to 359 million years ago, while the Upper Carboniferous is called the Pennsylvanian and ranges from 299 to 318 million years ago (fig. 2.1A).

Sedimentary rocks show the geologic column directly. Such rocks form from sediment (sand, mud, gravel) that was carried by water or air and eventually sank to the bottom of a body of water, where it accumulated and solidified over a long time.[3] Imagine finding a rock wall that includes sedimentary rock laid down during the Carboniferous Period. You would know just by looking at the wall that the older rocks are near the bottom and the younger ones are near the top. This is in fact a "rule" of sedimentation, the law of superposition: older sedimentary rocks are lower and younger rocks are higher. For this reason, the terms "early" and "late" are sometimes used instead of "lower" and "upper." For example, the Mississippian Period can be called the Early Carboniferous and the Pennsylvanian Period can be called the Late Carboniferous. Some sedimentary rocks like salt beds or coral reefs do not form by the transport and accumulation of particles, but they still obey the law of superposition.

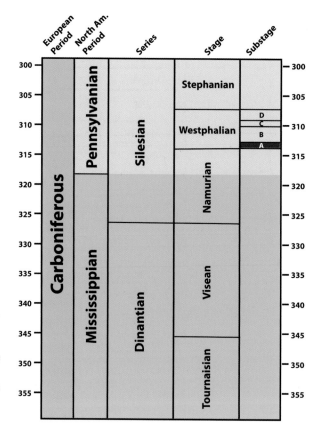

FIGURE 2.1*A* Geologic time scale of the Carboniferous Period, showing the Westphalian A (red horizontal band at right), the geologic time window during which the sedimentary rocks at the Union Chapel Mine were deposited. Times are in millions of years.

Alabama's swamp forests flourished during much of the early Pennsylvanian, but the trackways are restricted to a relatively thin interval of rock that formed during a narrow window of geologic time called the Westphalian A stage (red bar in fig. 2.1*A*).[4] The Westphalian is named after Westphalia, Germany, where rock layers of this age were first recognized and described. The Westphalian A is part of the Pennsylvanian Period, which itself is part of the Paleozoic Era, or "era of ancient life" (fig. 2.1*B*). The Paleozoic is older than the more familiar Mesozoic Era, or "era of middle life," commonly known as the Age of Dinosaurs. To encounter Mesozoic rocks in Alabama, you have to go south of Tuscaloosa. Mesozoic exposures in Alabama commonly involve chalk deposits and marine fossils such as shells and shark teeth. Only late (Cretaceous) Mesozoic rocks are exposed in Alabama; rocks from the older Triassic

EON	ERA	PERIOD	EPOCH		Ma
PHANEROZOIC	CENOZOIC	QUATERNARY	HOLOCENE		0.011
			PLEISTOCENE	Upper	0.8
				Lower	2.4
		TERTIARY — NEOGENE	PLIOCENE	Upper	3.6
				Lower	5.3
			MIOCENE	Upper	11.2
				Middle	16.4
				Lower	23.0
		TERTIARY — PALEOGENE	OLIGOCENE	Upper	28.5
				Lower	34.0
			EOCENE	Upper	41.3
				Middle	49.0
				Lower	55.8
			PALEOCENE	Upper	61.0
				Lower	65.5
	MESOZOIC	CRETACEOUS	Upper		99.6
			Lower		145
		JURASSIC	Upper		161
			Middle		176
			Lower		200
		TRIASSIC	Upper		228
			Middle		245
			Lower		251
	PALEOZOIC	PERMIAN	Upper		260
			Middle		271
			Lower		299
		PENNSYLVANIAN	Upper		306
			Middle		311 *
			Lower		318
		MISSISSIPPIAN	Upper		326
			Middle		345
			Lower		359
		DEVONIAN	Upper		385
			Middle		397
			Lower		416
		SILURIAN	Upper		419
			Lower		423
		ORDOVICIAN	Upper		428
			Middle		444
			Lower		488
		CAMBRIAN	Upper		501
			Middle		513
			Lower		542
PRECAMBRIAN	PROTEROZOIC	NEOPROTEROZOIC			1000
		MESOPROTEROZOIC			1600
		PALEOPROTEROZOIC			2500
	ARCHEAN	Upper			3200
		Lower			4000
	HADEAN				

FIGURE 2.1*B* Complete geologic time scale showing where the Carboniferous Period (indicated by the Pennsylvanian and Mississippian Periods together) fits into the Paleozoic Era and within the other eras of Earth's history. The asterisk shows where the Union Chapel Mine lies in the time sequence.

and Jurassic Periods are well below the surface. Most Cretaceous surface rocks in Alabama are marine (meaning they formed at the bottom of a salty sea), which is why fossils of dinosaurs are not common in the state. Still, some interesting species, such as the carnivorous theropod *Appalachiosaurus montgomeriensis*, have been found.[5]

The current era (the one we live in) is called the Cenozoic, or "era of recent life." This era began 65 million years ago with the sudden extinction of the dinosaurs and many other groups, and it is commonly known as the Age of Mammals. The three main eras of "visible life": Paleozoic, Mesozoic, and Cenozoic, together define the 542 million-year eon called the Phanerozoic.

One of the strangest things about coal-age Walker County is that instead of being at 34° north latitude, where it is now, it was south of the equator and well within the tropics. How do we know this? Apart from the geologic evidence, the kinds of plants that left fossils in Alabama coal mines were those that thrived in a swampy, tropical environment. Without such an environment, with its shallow, oxygen-poor water, there could be no coal. If we could have photographed the Earth from space at the time, we would not have seen the familiar continents of our time. Instead, we would have seen a supercontinent, in which all the major continents were merged into one. We know that a big collision of continents occurred in the southeastern United States because it left marks that are recognizable today. Instead of the dry land that characterizes Alabama, Mississippi, and Tennessee now, our Pennsylvanian space photograph would show a huge inland sea near a very high mountain range. The sea formed because the mountains had formed. The geologic name for the sea basin is the Black Warrior Basin, and while the sea disappeared a long time ago, the nearby mountains can still be seen: we call them the Appalachians.[6]

Coal mining in Alabama is challenging work. Coal beds are not generally exposed directly on the Earth's surface where they can be excavated with little effort or expense. An outcrop may be seen that has to be followed underground. As long as a seam is not too far below the surface, say less than 60 m (200 ft), it is most economical and probably safest to excavate the seam by removing the rock and soil (overburden) that covers it. This is called surface or "strip" mining. With such an approach,

Mary Lee

Blue Creek

Jagger

~6m

Crescent Valley Mine

seams can be followed wherever they go until they are too deep to be extracted or have been exhausted.

Figure 2.2*A* shows a typical vertical cliff, or highwall, created by strip-mining. Three coal seams are exposed in this case, historically named the Mary Lee, the Blue Creek, and the Jagger, each separated from the next by about 6 to 7 m (20 ft). A large swath of land had to be removed to expose a wall like this. In the excavated areas, the overburden of each seam was removed and the coal was extracted in the form of small chunks. Figure 2.2*B* shows coal mined from the narrow Blue Creek seam and gathered into a large pile. The coal is bituminous (made of a tarlike substance called bitumen)[7] and is in the form of dark blocky chunks, often with a banded appearance.

As effective as strip-mining is, the movement of tons of rock and the creation of highwalls can greatly change the landscape. The process of

FIGURE 2.2 *A* A highwall at the Crescent Valley Mine in western Walker County, showing three labeled coal seams. The seams are about 6–7 m (20 ft) apart. The large "caves" are areas where Jagger coal was extracted decades ago.

FIGURE 2.2*B* Mining Blue Creek coal at the Kansas Mine No. 2 near Eldridge, in western Walker County. The blocky character of the pieces is typical of bituminous coal.

extracting coal can release unwanted contaminants that, depending on mine location, can affect local water supplies, sometimes fueling vocal opposition to the mining. Strip-mining may also involve blasting, which can disturb local residents, and is inherently disruptive of surface landscapes in comparison with underground mining.

Stratigraphy is the science of reading the sedimentary record, and of knowing how to link sedimentary rocks to different periods of Earth's history and to relate rock sequences in different parts of the world to one another. The highwall of a typical surface coal mine represents actual deposits laid down during the Carboniferous Period. In such a wall, we see layer upon layer (strata) of sedimentary rocks stacked in a regular time sequence. In these rocks is preserved a record of past events, including the periods of intense forest growth that led to formation of the coal beds. Although each bed formed in much the same way, the quality (purity) of any given part of a seam can depend on how much mud was brought into the swamp by flooding rivers. The less mud brought in, the purer the coal, so the best coal was usually deposited some distance away from active river channels. Not every bit of mud, sand, or plant debris is preserved for posterity. The highwall may conceal more than it shows, and for every delicate shale layer you see, many others were eroded by floods while they were still soft mud.

The sedimentary rocks in Walker County are part of what geologists call the Pottsville Formation, so named because the rocks were first recognized in North America near Pottsville, Pennsylvania. The formation is more than a mile thick in Alabama, and close inspection of the strata reveals a periodicity in the way the sediments were deposited (chapter 9). All the layers of Pottsville rock are thought to have been deposited on a time scale of just a few million years. Gazing into the ancient layers of rock exposed in a Pottsville highwall can be a humbling experience. The layers are like a photo album that has not been opened for millions of years. When sunlight last shone on the mud that became the strata now exposed, the world was a very different place. Dinosaurs were yet to evolve, the Mesozoic was far in the future, and human beings were not even imaginable. As we show in chapter 6, even the night sky would have been unrecognizable.

It is in the shaley strata between coal seams that fossils are most abundant. The plant fossils in a coal mine lie sandwiched between rock layers that are broken up by the earth-moving equipment used in mining operations. Usually, these fossils are highly compressed and sometimes carbonized, meaning that the remains of the original plant are present in a coalified form. Although coal itself is preserved plant material, the plants making it up have been so drastically changed by chemical processes and external forces that few people would include pieces of coal in a fossil collection. Fossils of seed ferns (an extinct group of plants superficially similar to modern ferns) and their seed cases and pollen organs, as well as impressions or even casts of the bark, branches, and roots of magnificent ancient trees, are found at most surface mines in Walker County. The plant fossils at a mine can tell us a great deal about the environment where the plants lived.

In addition to the distinctive types of plants, the majestic forests of coal-age Alabama also teemed with animal life, including aquatic creatures that lived offshore from coastal swamp forests and in the bayous and ponds within the swamps. Extensive mudflats were where terrestrial animals hunted, and where some aquatic animals came to breed. Bones and other body parts of these animals would have been destroyed by the acidic mudflat soil, making it difficult for us to know what the animals actually looked like. The animals that lived near the ancient swamps would nevertheless leave in the wet mud traces of their daily activities that, under the right circumstances, could be preserved in sediment that solidified over time. These traces can be exposed in the layers of rock broken up by mining operations, just as plant fossils are. Trackways are just one of the most important kinds of trace fossil. Any animal will leave a distinctive pattern of footprints when it walks out onto wet mud. Activities such as walking, crawling, jumping, swimming, burrowing, and even "resting" can leave traces for us to find.

Because trackways record what a *living* animal was doing at a given instant, they are without doubt the most fascinating kind of fossil exposed by coal mining. They reveal activity that no other fossil can represent. They are totally about *life*. Trackways and other trace fossils record what primitive animals (reptiles, amphibians, fish, insects, and more) were doing out on the mudflat, however mundane the activity. When you see a trackway crossing a slab, you know that an animal once walked across that surface when it was mud. Bones can tell us what an animal looked like, but trackways can tell us how an animal lived.

Coal mining in Alabama did not end in the early twentieth century but continues at present, and tracks like those described by Aldrich and Jones can still be found. The No. 11 Mine of the Galloway Coal Company was simply the first mine where people noticed the tracks. Many specimens were probably seen over the following years at other mines in Walker County but were neither systematically collected nor documented beyond what had already been done in 1930. Yet the potential for further knowledge exists at almost any mine in Walker County. Each mine could sample a different environment or a different coal seam. These differences would affect the kinds of fossils found. For example, while plants are abundant at virtually all Walker County mines, different plant fossils are found in different places, recording ancient environments ranging from bayou to hilltop. Also, tracks may or may not be abundant in a given mine. Integrating all this information could provide a broad window into life in coal-age Alabama.

While fossil trackway research was minimal in Alabama from 1930 to 1999, things changed considerably when Ashley Allen and his fellow amateur collectors in the BPS discovered and documented the Union Chapel Mine in eastern Walker County. In a matter of months, the ancient footprints of Walker County would suddenly come back into the limelight and garner more interest than they had in the previous seventy years. Later in this book, the unique series of events that brought the ancient footprints of Walker County back into public consciousness will be described. Let us first introduce the reader to some of the terminology of trace fossils.

3

Portals to Deep Time

Fossils are the preserved remains or traces of ancient organisms. The vast majority of organisms that ever lived left no fossils, because fossilization is a very rare occurrence. The preservation of organic remains requires a special sequence of events and circumstances that vary depending on what is being preserved. Fossils are like portals to deep time: they can take us back so far in time that the world was truly a different place when the animals that left the fossils were alive. In general, the Age of Humankind (the last three to four million years) and the world of fossilized animals or plants hardly overlap at all.

The previous chapters showed the draw that fossils can exert on people. Alabama state geologist Walter B. Jones believed that the tracks found near Carbon Hill were scientifically important enough to merit a rapid publication, and he wrote elegantly of their discovery. He saw to it that he and his staff collected enough specimens to document the find well enough for future studies to build upon. High school science teacher Ashley Allen saw coal-age fossils as a way of teaching his class how science works and inspiring his students to learn more about the natural world around them. His enthusiasm for the coal age, and especially for Paleozoic footprints, had been encouraged by a geology teacher he deeply respected. This led to the biggest discovery of Allen's life.

Professional researchers see fossils as clues to what the world was like in the remote past, and they enjoy the detective work needed to understand ancient plants and animals based on the often meager (and also usually severely biased) evidence nature provides. Professional paleontology is difficult but very rewarding work that can lead to recognition of new species and new views of past ecosystems. Amateur collectors see fossils as relics of bygone eras and enjoy mostly the search for fossils, their beauty and mystique, and the often out-of-the-way places people must go to find them. There are many more amateur fossil collectors than there are professional paleontologists, but both can share a similar level of passion for the subject. Who, for example, came through adolescence without loving dinosaurs?

In this chapter, we take a look at the nature of fossils, how they form, and the different types of fossils that are found. This is essential in order to understand the fossils found at the Minkin Paleozoic Footprint Site.

Two Kinds of Fossils

When an organism dies it can leave behind a variety of evidence that it once existed. Take Thomas Jefferson, for example. Jefferson had children, grandchildren, and so on. He contributed to the ongoing evolution of the human race by propagating his particular set of genes. He also left behind 206 bones, which were very properly laid to rest in his grave. If these bones were examined, one could learn a lot about his physical appearance, such as his height, build, and facial structure. But the best way to learn about Jefferson is to study his works. Jefferson left behind a beautiful house, Monticello, in which he designed many ingenious appliances, such as one of the world's first dumbwaiters. He also left behind a great deal of writing. The Declaration of Independence comes to mind. These are in a way traces of Jefferson's former existence, but they are not part of Jefferson himself.

An amphibian or reptile living near a coal-age swamp forest would have had the opportunity to leave the same three kinds of evidence. It could leave a skeleton. No tetrapod skeletons have yet been collected from the coal-age strata of Alabama, although skeletons of similar age have been found in other parts of the world (for example, fig. 3.1, top; Berman et al. 2004). The animal could leave descendants, which we might be able to identify using fossil bones or preserved genetic material. But while we do not have such material for the coal-age animals of Alabama, we do have their "works." We have thousands of footprints, tail-drag marks, and other trace fossils that were left in muddy areas (fig. 3.1, bottom). We know how big the animals were, how fast they

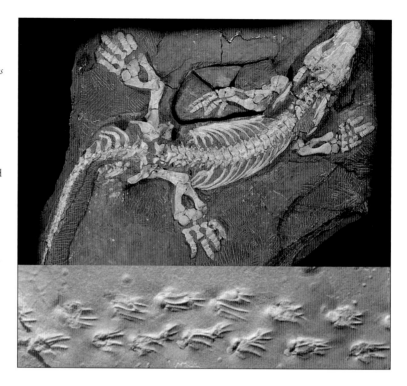

FIGURE 3.1 *Top, Orobates pabsti,* body fossil of a Paleozoic reptile-like amphibian (Permian age) housed in the Museum der Natur, Gotha, Germany. The animal was an early herbivorous tetrapod (Berman et al. 2004), approximately 1 m (3 ft) long from tip of tail to snout with a skull length of 12 cm (4.7 in). *Bottom,* Union Chapel Mine trace fossil UCM 109, showing the well-defined tracks of a small amphibian, positive hyporelief (meaning a natural cast). Small burrow in upper left is *Treptichnus apsorum.*

moved, the consistency of sediment they preferred to walk through, whether they moved individually or in groups, and how many toes they had, among other things. We also know exactly where they lived, and what the environment was like. Bones can be washed from one place to another by floods, or they can be dragged by carnivores. Footprints do not move. If you step in the mud, anybody who sees and recognizes your footprints knows where you have been. Trace fossils and body fossils provide different kinds of information about the ancient world.

The preservation of body fossils, especially those of terrestrial vertebrates, is a most unlikely process. In general, only the hardest parts of an organism, like bones or teeth, will have a chance to fossilize. The key to fossilization is rapid burial, which can occur in violent storms (through, for example, flash floods) or by other means (such as volcanic eruptions). Rapid burial by sediment in a watery environment protects the organic material from scavenging, normal decay, and weathering. Instead of simply dissolving and leaving no trace of its former existence, organic material in this circumstance can be slowly replaced with minerals from its surroundings, producing a rock copy of the hard parts. Even the empty spaces in a bone can fill with minerals over time, a process geologists call permineralization.

The sediment itself will eventually solidify and form what is called the "matrix" that a body fossil is imbedded in. If a body fossil is exposed on the surface for a long time, it can weather out of its matrix naturally and be collected as found. Often, however, a body fossil is found still encased in its matrix. Removal of bone fossils from this matrix can be extremely delicate and difficult and, depending on how badly broken up the skeleton is, or on how complicated and large it is, could take years of tedious work. People who engage in the slow process of extracting body fossils from matrix are called preparators. For some body fossils, especially those of relatively small animals with delicate shells or bones, a skeleton may best be left in its matrix and studied as preserved.

Like bones or other hard organic material, footprints or other traces made in a soft substrate (like wet mud) also depend on rapid burial in order to be preserved as trace fossils. But in contrast to body fossils, trace fossils do not require "preparation" beyond cleaning because what is preserved is not an organism itself, just the fleeting impression of an organism's movement in an area. This impression is simply preserved as solidified mud, encased only in the sense that it was covered by another layer of sediment. Since a single animal can leave many footprints or other kinds of traces, trace fossils can be much more abundant than body fossils, and they can also be more informative. If there are no body fossils, as in the case of the Carboniferous rocks of Alabama, then the *only information* we might have about the former existence of a particular organism is derived from the trace fossils it created. Trace fossils can provide so much information about behavior and ecological relationships that they overshadow what can be learned about morphology and evolutionary relationships from the few body fossils that have been found elsewhere. A typical trackway, for example (such as UCM 109, fig. 3.1, bottom), can reveal information on an animal's gait (the way it walks), its association with other organisms, and some aspects of its foot morphology. It will be very exciting if amphibian or reptile body fossils are ever found in Carboniferous strata in Alabama, but bones

FIGURE 3.2 Cast fossil of a small lycopod trunk, probably of the genus *Sigillaria*. The vertical dark bands are the carbonized (coalified) remains of the original plant. The base diameter of the cast is about 23 cm (9 in). Known as the "Devils Tower" specimen because of its resemblance to Devils Tower National Monument in Wyoming, the fossil was collected by Ron Buta from the Union Chapel Mine in 2000.

would have to be numerous and well preserved to provide more insight into the coal-age communities of the southeastern United States than is already known or discoverable from trace fossils.[1]

There are other fossils that can provide information like that provided by body fossils. These are called *casts*. A cast is a solidified three-dimensional sediment impression of an organism. A typical cast found at a coal mine is a lycopod tree trunk, which in life had softer tissue inside the trunk. The soft tissue decayed and the hollow trunk filled with mud or sand that eventually solidified. Surrounding the cast may be carbonized plant material, the remains of the bark of the tree (fig. 3.2). Also commonly found in coal mines are casts of smaller coal-age plants called horsetails, whose fossils are named *Calamites*. A third type of cast found in coal mines is *Stigmaria*, representing the roots of ancient lycopods. Although not yet found in Alabama coal mines, casts of the bodies of Carboniferous amphibians are known from other parts of the world.

Note that a cast of a tree trunk is not the same as petrified wood. The former is made of the same material as the sedimentary rock layers and preserves no details of the cell structure of the original plant, while the latter forms via total replacement of the plant tissue by minerals that may preserve fine details of its original organic structure.

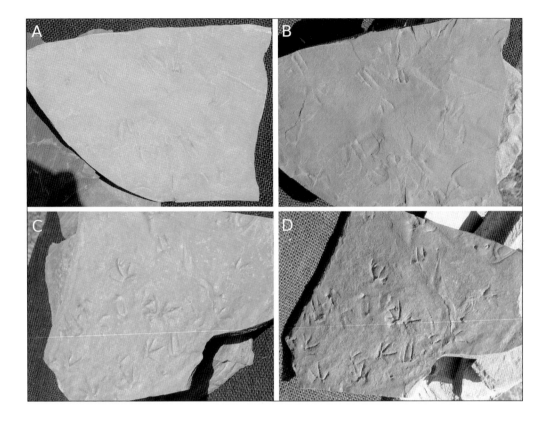

SHADOW AND LIGHT

Because trace fossils are nothing more than shapes or distinct patterns in a rock, there may be very little difference between the trace fossil and the rock in terms of color, composition, or texture. Body fossils may be immediately obvious when they are exposed, because bone can have a different color or texture from the surrounding rock matrix. Trace fossils are much more about light and shadow.

Weather and angle of incident light are important factors in seeing trackways. Trackways are most easily noticed on sunny days when the sun is fairly low in the sky, or when a slab is tipped at a high angle to the sun. Figure 3.3 illustrates the effect of sun angle on two track specimens of *Cincosaurus cobbi* (chapter 15) from the Crescent Valley Mine in Walker County. (The naming of species is explained below in the

FIGURE 3.3 Effects of sun angle for recognizing tracks in the field. When the sun's angle is high (*A* and *C*), tracks are more difficult to see even if strongly impressed. When the sun's angle is low (*B* and *D*), tracks are easily seen even if weakly impressed. These *Cincosaurus cobbi* tracks are from the Crescent Valley Mine (chapter 19). *A* and *B* are of the specimen CVM 146, while *C* and *D* are of the specimen CVM 147.

section titled "What's in a Name?"). Figure 3.3*A* shows a light trackway that would easily be missed when the sun's angle is high, but which is obvious when the sun's angle is low (fig. 3.3*B*). Similarly, the trackway shown in figure 3.3*C* is visible fairly easily at a high sun angle but is dramatically better seen at a low sun angle (fig. 3.3*D*).

On rainy, wet days, trace fossils can be seen on rocks that would be featureless on cloudy, dry days. For example, on a BPS field trip on May 28, 2000, a cloudy, rainy day, coauthor Buta found a set of tracks made by a small amphibian that appeared to have walked around something in its path (chapter 8). In the field trip report posted on the BPS website at the time, the find is recounted this way: "I noticed these tracks after splitting a rock. The two halves, giving direct and inverse impressions, were wet and the tracks obvious. However, after the pieces dried in the back of my truck, I could not see the tracks!"[2] In spite of the difficulties associated with high sun angle or a dry, cloudy day, deep tracks will still be obvious. Only shallow, lighter tracks would be missed under such conditions.

Although trackways were found over most of the (original) eighty-acre UCM, some areas were richer than others. When the New Acton Coal Company set about getting coal out of the ground, the overburden was broken up and piled out of the way. These spoil piles were made as close as possible to the source of the broken rock. If there were more trackways in some places than others before mining began, the original distribution was preserved in the spoil piles. Clues to the original environmental conditions on the ancient muddy terrain (high versus low areas, wet versus dry areas, vegetated versus barren, for example) can be discerned from the spoil piles. Much more could have been learned if the samples had been found in place (that is, in their actual rock layer position, or "in situ"). For a surface coal mine, such collecting would be a delicate, dangerous, and potentially expensive operation that has not been tried. Chapter 18 discusses some of the factors affecting in situ trace-fossil collecting in coal-age rocks.

Success in fossil collecting often depends on knowing how to recognize the right rocks. At the UCM, the most productive rock for finding vertebrate trackways was thinly bedded shale with shiny surfaces, usually with layers no more than two to three millimeters thick. The best-preserved tracks would have formed in areas where the sediment consisted of very fine particles deposited in very slow-moving water (P. Atkinson, pers. comm.).

Before partial mine reclamation, which took place in December 2000, fossil hunting at the UCM involved clambering over unstable rock piles in search of patterns of light and dark on the rock surfaces. After reclamation had been partly accomplished, the effect of light was unchanged, and the rock piles were not as dangerous. Normally, mine reclamation yields a grassy surface that completely conceals the rock that was exposed by mining. The UCM was not fully reclaimed because the BPS was actively trying to get a waiver during the period when reclamation was supposed to take place. Partial reclamation of the mine did make the place safer, but the intensely fractured vertical highwall remains a potential hazard.

There are several ways of finding trackways among the spoil piles of a surface coal mine. A slab with tracks on one side may be found with the tracks lying face-up and exposed on the surface, or the slab may be lying face-down so that the tracks are seen only when it is turned over. In either situation, unless the slab is very heavy, collecting the trackway is straightforward provided it can be seen easily in the prevailing light. A more typical situation is that a slab has a trackway sandwiched inside, such that to expose the trace, it is necessary to split the slab along the track-bearing horizon. A clean split will give two trackways: the actual impression in the rock, and the natural cast formed by sediment that covered that impression. The natural cast is called a counterimpression and will consist of elevated marks rather than depressions.

Although many specimens found at the UCM were isolated trackways or pairs of impressions and counterimpressions, occasionally a single large slab yielded its own mother lode. In May 2003, a large rock was found with several bedding planes densely covered with well-preserved invertebrate trackways. Initially, on one bedding plane this rock yielded a sequence of four large tracks identified as *Attenosaurus subulensis*, tracks inferred to have been made by an alligator-sized protoreptile, or reptile-like amphibian (chapter 15). To see what else might be in a rock like this, it was helpful to stand it on its side so it could be split more easily. When lifted (requiring three people), the slab split, leaving some pieces from the underside covered with invertebrate trackways that sparkled in full sunlight. When the remainder of the slab was split, even more densely covered bedding planes were exposed. Especially remarkable was the extreme regularity of many of the trackways. This was a case in which most of the good stuff was near the bottom of a massive slab, which might explain why the rock had received little attention previously (Haubold et al. 2005b, plate 21).[3]

TRACK FORMATION

Animals that walk on wet mud will leave tracks no matter where that mud is located. Figure 3.4 shows modern deer tracks left in brown mud on Mount Cheaha in east-central Alabama. As the animals walked on the mud, their feet left regular impressions that preserved some aspects of their foot morphology, such as the two-toed nature of the hooves. In mud of a certain consistency, the track remained clean even after the animal lifted its foot, as in the tracks at the left in figure 3.4, which show two clear depressions and readily identify the tracks as those of deer. However, most of the tracks in figure 3.4 are little more than round depressions where the hoof details can barely be seen. If some parts of a muddy area are drier than others, or if tracks are made at different times in the same area—for example, when it is very wet and then after it has some time to dry—this affects how well defined the footprints will be. Dry mud will not take footprints at all. If an animal walks from an area covered with a shallow film of water onto a drier area, the tracks will likely become fainter and may even disappear. Presumably, the Mount Cheaha mud is visited frequently by deer at times of different degrees of wetness, and this could account for the way some of the tracks look. These modern tracks show us the difficulties we face when we try to interpret what trackways in the fossil record might signify. A footprint is shaped not just by the foot but by the consistency of the mud the footprint is impressed in, and also by the way the animal is walking.

Because trace fossils are the preserved effects of the behavior of organisms in an environmental context, in principle, the deer tracks on Mount Cheaha could be preserved as trace fossils, but only if they were buried quickly and deeply enough so that they did not wash or erode away. Since the tracks were found near the summit of Mount Cheaha,

there is little chance they will be buried. Their preservation potential is essentially zero, since they will be continually exposed to weather and erosion. Thus, to preserve a trace of an organism's behavior, such as a deer walking on mud on Mount Cheaha, very special conditions are needed. Most of the time, the right conditions will not be present and the traces will not be preserved.

All footprints form the same way. A foot pushes down into sediment, compacting what is immediately beneath it and pushing some up as a rim around the foot. Then, many animals push off as they lift their feet. The dynamics of this pushing off make the footprints deeper toward the front of the animal. Some footprints show a ridge of sediment pushed up right behind. Soupy sediment is pulled up behind the feet as they emerge from the muck and then falls back down, which may result in complex shapes. Walking always involves a rhythmic, cyclic motion of

FIGURE 3.4 Deer tracks left in exposed soil on Mount Cheaha in east-central Alabama, March 2012. The tracks do not all look alike, indicating that the soil was stepped in at different degrees of wetness, and also that multiple animals of somewhat different foot size walked in this area.

feet, such that right and left footprints alternate along the trackway on either side of the midline.

Other types of traces are made by different modes of locomotion such as crawling and hopping (jumping) and can form differently. Jumps or hops can be short or long, which can leave different-looking traces, as we show in chapter 16. Burrows can be made by wormlike organisms ingesting sediment at the front and excreting it at the rear, such that the burrows are always mostly full of fecal (waste) pellets. Other burrowing organisms use their legs to push sediment behind them as they go. These burrows are also mostly full, and the fill material may resemble fecal pellets. Still other burrowers use water circulating in open burrows to provide themselves with oxygen or food particles. Some animals lurk in open burrows and wait for prey to happen by. Many kinds of burrows have diagnostic patterns of sediment packing or other characteristics that help us recognize their makers or the behaviors that produced them. Most Minkin Site burrows appear to have been made by fly larvae or other stubby, cylindrical insect larvae (Rindsberg and Kopaska-Merkel 2005).

How do we know which way was "up" when we find a trace fossil? Trace fossils may be found on top of, within, or on the bottom of a layer of sedimentary rock. They may be raised above the surface of the surrounding rock or they may consist of depressions in the rock surface. Burrows may penetrate right through a rock. Because we know that footprints originate in depressions made in the sediment surface, a piece of shale with a row of footprints pressed into the top ("negative epirelief") is right side up. This means that any unfamiliar trace fossils in that piece of rock are also right side up. By contrast, a slab on which trackways are elevated relative to their surroundings ("positive hyporelief") is upside down.

Undertracks versus Surface Tracks

When an animal places a foot in wet mud, it does not make just one footprint. The surface mud may be too fluid to register an accurate foot impression, or *surface track*, but a footprint can still be preserved because the weight of the animal causes its feet to compress the underlying sediment, leaving impressions of the foot even on layers that were not originally in contact with the foot. The heavier the animal, the deeper its influence is felt. Each affected layer can yield a trace fossil, an *undertrack* (or "ghost print"; Lockley and Hunt 1995), in which successively deeper layers preserve less and less information about the shape of the foot and how it was placed and moved. Quite different trackways, formed at varying depths below the sediment surface, can be made by animals of a single species engaging in the same kind of behavior. Surface tracks are much more vulnerable than undertracks to being washed away even by slow water movement. The preservation potential for surface tracks is therefore generally very low compared to undertracks. As Seilacher (2007) notes, undertrack impressions are *already buried* and can be well preserved even if they are only a few millimeters below the sediment surface. Well-defined surface tracks are also referred to as primary tracks. The deer tracks in figure 3.4 are primary tracks.

Figure 3.5 shows a schematic that illustrates the relationship between impressions and counterimpressions, as well as surface tracks and

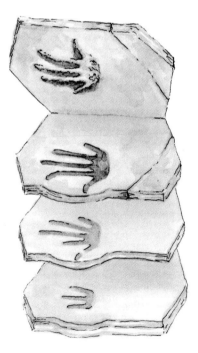

surface track natural cast

surface track

shallow undertack

deeper undertrack

FIGURE 3.5 Schematic of the relationships between impressions and counterimpressions, and between surface tracks and undertracks. A counterimpression is a raised natural cast of the original impression of an animal's foot, formed when fresh sediment fills an existing footprint and later hardens to rock. Undertracks are foot impressions carried into distorted lower layers. The deeper the undertrack, the less information on foot morphology is preserved. Not shown in the illustration are the natural casts of the undertracks that would be seen in the layers immediately above.

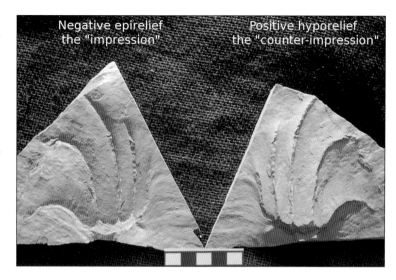

FIGURE 3.6 Example of an impression (CVM 209, left, negative epirelief) and counterimpression (CVM 210, right, positive hyporelief) of a large *Attenosaurus subulensis* track. A single rock was split along a plane to reveal both halves of a single preserved footprint. The impression is right side up, while the counterimpression is upside down. The specimen shows all five digits and may be a surface or near-surface track. Collected from the Crescent Valley Mine.

undertracks. A tetrapod track is exposed as a set of toe impressions in negative epirelief in a rock layer that was once a primary surface. The track was exposed by splitting the rock, which also revealed the counterimpression, the natural cast of the footprint in positive hyporelief. (A real example is shown in figure 3.6.) In the area where the foot was in contact with the surface, the layers below are distorted and toe impressions are transmitted to deeper layers. However, even though the actual foot had five toes, only three are seen preserved in the undertrack.

Loss of digits is one of the main features of undertracks. What the schematic shows is illustrated by the pictures in figure 3.7A–D. These show two fossil tracks of *Attenosaurus subulensis*, made by a large animal that was heavy enough to leave deep undertracks. The tracks were found at the Crescent Valley Mine (chapter 19) in a large, heavy slab that was upside down, and the counterimpressions, or natural casts of the tracks, survived the splitting of the rock better than did the impressions. Figure 3.7A shows the tracks that were closer to the original surface. The right track is a forefoot impression (also known as a manus), while the left one is a hindfoot impression (also known as a pes). In figure 3.7B, a layer of rock 5.5 mm thick is put into place over the tracks in figure 3.7A to reveal part of an undertrack. This shows a significant lateral displacement between the toes of the undertrack and those of the near-surface

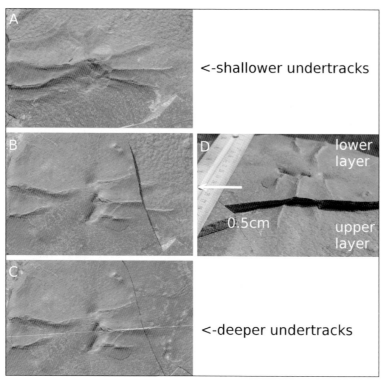

<-shallower undertracks

lower layer

0.5cm

upper layer

<-deeper undertracks

FIGURE 3.7 Crescent Valley Mine specimen CVM 650 showing how undertracks change with depth below the surface. The slab is upside-down and shows counterimpressions. (*A*) Two *Attenosaurus subulensis* tracks, the right one a forefoot (manus) and the left one a hindfoot (pes). These are near-surface undertracks. (*B*) The same two tracks, with part of the hyporelief surface of a layer 5.5 mm lower put in place to show how a deeper undertrack compares with the near-surface track. (*C*) The complete undertrack can be compared with the near-surface track shown in *A*. (*D*) A side view of *B* to show the mismatch between the near-surface track and the undertrack. The arrow points to the perspective in *B*.

track. In addition to the foot moving downward into the sediment, there was also a component of sideways motion from the body (best shown in the side view of fig. 3.7*D*). As the animal walked, its foot did not only go vertically into the sediment but pushed outward as well. When the remainder of the undertrack is put on (fig. 3.7*C*), we also see that the toes in the undertrack manus cast are shorter than those in the near-surface manus cast. This specimen shows that undertracks give the false impression that an animal's feet had fewer digits, shorter toes, and a wider extent than they actually did.

Figure 8.10 (chapter 8) shows an invertebrate example of the relationship between undertracks and likely surface tracks. UCM 1070 and 1071 are from opposite sides of a slab only a few millimeters thick and show beautiful traces made by a small horseshoe crab. The conspicuous telson (tail) drag mark and the complex patterns on either side suggest that UCM 1070 is a surface trace or a very shallow undertrace. The

FIGURE 3.8 Trackway specimen UCM 1797 dramatically shows how indistinct a surface track can be compared to an undertrack. The surface track lies only a millimeter or so above the undertrack. The undertrack shows clear multitoed footprints, but the surface track shows only a washed-out area with few toe prints visible. Two small pieces of rock were removed to reveal the undertrack on the left. Washed-out surface tracks form in soupy mud. Negative epirelief, *Nanopus reidiae.*

trackway in UCM 1071 is only a few millimeters below the UCM 1070 bedding plane. UCM 1071 is clearly the same trackway as UCM 1070, but UCM 1071 shows less detail because it is a deeper undertrack. All that is preserved in UCM 1071 is a biramous (double-branched) pattern of small *Y*s that represent the tips of the main legs of the horseshoe crab. Farther below the surface track, only the inner pattern of parallel *Y*s would be visible, a pattern that was found in one of Ashley Allen's discovery specimens (fig. 1.2, middle).

In both of these examples, the surface or near-surface trackway was not completely washed away. Figure 3.8 shows an example, UCM 1797, where the original surface trackway is barely visible. In this case, the surface trackway was made in mud too soft to take an impression, but on the left side of the picture we see a slightly lower layer where undertracks of the same trackway show well-defined multitoed footprints. The surface trackway is roughly horizontal and goes between the two small bumps, finally curving slightly downward to the right. The undertracks were exposed because a layer of the rock, at most a millimeter thick, had been weathered off.

Other examples of undertrace effects include tetrapod trackways where only the fore- or hindfoot tracks are seen. This can occur if one set of feet (say the forefeet) distorts slightly deeper layers than does the other set (the hind feet), which reflects the way an animal distributes its mass as it walks. If one of these deeper layers is exposed, then the other set of feet will either be very faint or not seen at all in the preserved trace fossil. An interesting example of of this phenomenon with respect to Mesozoic sauropod footprints was described by Lockley and Hunt (1995). A trackway showed mainly manus prints, which had been interpreted as being made when a large animal was swimming and touching the bottom with only its front feet. Lockley and Hunt argued that the lack of pes prints likely meant only that the animal put slightly more of its weight on its forefeet, so that these were impressed into deeper layers.

A final issue is that, depending on the substrate, trace fossils can show "extramorphological characters," which are morphological features tied more to the properties of the substrate and the way the animal was moving in it than to foot shape or other intrinsic characteristics. Anything that is not a near-faithful mold of the animal's foot morphology is extramorphological to some degree. The composition and moisture content of sediment determines how deeply footprints are impressed and in what way they preserve the shape of the organism's foot and the direction and configuration of the step. Trackways fade away or become indistinct where animals walked from wet mud to drier mud, as we have already noted. An example of this is shown in figure 19.7.

The classification and recognition of trace fossils must allow for substrate conditions and the different possible tracks that a single type of animal can make. *Most trace fossils from Walker County are undertraces.* Undertraces are more common because they have a greater preservation potential than surface traces. The dearth of tetrapod surface tracks suggests that the animals were walking in very fluid mud that could not, in general, hold the traces.

What Do Footprints Tell Us?

In an excellent guide to fossil footprints, Lockley and Peterson (2002, 25–26, 74) summarized the basic things we can learn from such traces, even if no body fossils of the trackmakers are ever found. In general, tracks are most likely to be made in wet ground, which favors finding them in sedimentary rock layers representing ancient shorelines. Thus, when the trackmakers were alive, the area where they left their footprints was open, relatively flat, and easily accessible by walking. But fossil tracks may be so old that the land has changed since the time they were made. For example, New Mexico paleontologist Jerry MacDonald (chapter 10) found trackways of ancient pelycosaurs (sail-backed mammal-like reptiles) that could be followed directly into the solid rock of a mountain. Clearly, the mountain did not exist when the trackmaker was alive. The tracks were buried, uplifted during an episode of mountain building, and then exposed later by erosion. Lockley and Peterson showed a spectacular trackway made by a Cretaceous sauropod on what is now a *vertical cliff face* in Bolivia. The trackway is hundreds of feet long. Geological forces pushed up vertically what was once a flat, nearly horizontal shoreline.

Seilacher (2007) summarizes three conditions needed to preserve footprints as fossil traces: (1) the mud must be cohesive enough to hold the trackway together after the animal has lifted its foot up; (2) something must be able to act as a casting agent; and (3) the casting must be completed before the footprint erodes away. Lockley and Peterson described two ways these conditions might be satisfied. The first is that an animal leaves a footprint in wet mud that has a chance to dry a little before the next flooding (which could be periodic). Then when the flooding occurs the track will not necessarily be washed away completely but can fill with sediment, preserving the footprint. This is called the "cover up" process and would probably work well on the Mount Cheaha deer tracks in figure 3.4. In the second process, a layer of firmer substrate below is covered by a softer, wetter sediment, such that when an animal lifts its foot out of the softer sediment, the impression it leaves in the harder sediment is covered immediately and can be preserved. This is called the "direct deposit" process.

Particularly important information encoded by tracks illustrates the manner in which the animals were moving. In the case of dinosaurs, trackways indicate that some traveled in herds. This is demonstrated by long sets of parallel tracks made by groups of similar animals. An excellent set of parallel sauropod tracks is shown by Lockley and Peterson (2002, 68). UCM vertebrates may also have occasionally traveled in

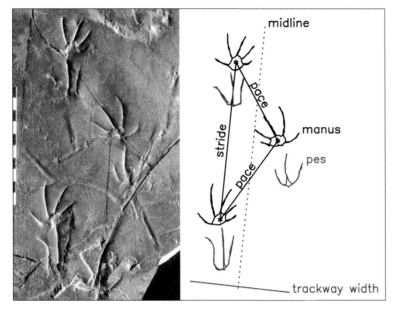

FIGURE 3.9 *Left*, UCM 263, a well-preserved undertrack of *Cincosaurus cobbi*, positive hyporelief; *right*, a schematic of UCM 263 illustrating standard terminology of tracks and trackways. The pace is the distance between a left footprint and the next right footprint. The stride is the distance between two successive left or right footprints. The trackway width (straddle) is the distance from one outer side of the trackway to the other side. The midline is the line over which the body was suspended.

groups (chapter 8). However, Lockley and Hunt (1995) caution that parallel trackways could mean nothing more than that an area was well traveled as a preferred route, such as a shoreline along an ancient lake.

Tracks provide other kinds of detailed information. Figure 3.9 shows a beautiful *Cincosaurus cobbi* trace (chapter 15), UCM 263, with both manus and pes prints. Although the tracks are well defined, they are undertracks (the pes prints show marks of only three toes, whereas the trackmaker had five). In the right panel of figure 3.9, some of the terminology of trackway morphology is illustrated using UCM 263 as a model. The midline is the symmetry line of the trackway. The step or pace is the distance between one footprint and the next (left to right or right to left). The stride is the distance between successive left or right footprints; in other words, two steps. The pace angulation is the angle between three successive footprints (such as the angle between the two pace lines in fig. 3.9), which can be used to deduce walking, running, or crawling speed. The trackway width (also called straddle) provides information on body width and degree of sprawl.

In addition to individual footprints or crawl marks, animals can leave other impressions such as a "tail drag," or tail trace (which can be left by both tetrapods and invertebrates like horseshoe crabs). Lockley and Peterson (2002) note that tail drags are usually seen in the traces of long-bodied, short-legged animals like amphibians and lizards, and they are very rare for larger animals like dinosaurs or mammals. Some apparent tail traces can also be "belly traces" (or pelvis impressions), made when the animal's belly or pelvis dragged through wet mud. Even heel, ankle, and skin impressions are possible. While footprints may leave undertracks, tail traces may not penetrate as deeply. Deeper undertracks of a surface trackway with a tail-drag mark may not show the tail impression, as well as not show all the actual digits of some of the feet.

Vertebrate versus Invertebrate

Morphologically, the trackways of vertebrate animals are distinct and generally easily recognizable. The feet of tetrapods have multiple toes and a pad. Their orientation relative to the trackway midline may vary, but it is the midline that tells us precisely which way the animal was going. In addition, vertebrate trackways will generally show a left-right alternating pattern of footprints, with a configuration that depends on gait and speed. This pattern is called a *lateral sequence walk*, and it involves the animal being on three legs at any given time as it walks.[4] An ordinary walking trackway, such as the one shown in figure 19.4, has nearly uniform track spacing on both sides. In some trackways, the pes print nearly overlaps the manus print, as in figure 3.9. This likely indicates a faster gait. Gait is the name for different modes of movement, such as running versus galloping. Different gaits mean different footprint patterns.

In contrast, invertebrate trackways will generally not show an alternating pattern of pairs of forefoot and hindfoot prints. Such animals tend to be smaller than vertebrates and have many more leg-like appendages than do tetrapods. The movements of such creatures can involve walking, crawling, jumping (hopping), or burrowing, and thus a significant variety of traces is possible. While land invertebrates can make undertracks by deforming lower layers, as vertebrates do, aquatic invertebrate undertracks can involve piercing the substrate.[5]

Arthropod trackways, made by insects, crustaceans, and spiders, can show alternating patterns, as in a scorpion trackway, or ladder-like symmetric patterns (Seilacher 2007). An excellent example of ladder-like trackways is provided by the genus *Stiaria*, found in abundance at the Minkin Site. The marks made by the appendages of the maker of *Stiaria* can be toe-like, but the trackway is distinct from any vertebrate trackway. The number of marks in a typical invertebrate trackway is tied to the number of legs and other features. It is also not necessarily obvious from the morphology of an arthropod trackway which way the animal was going. The many other possibilities of invertebrate trace fossils are beautifully illustrated by Seilacher (2007). Those found at the Minkin Site are described in chapter 16.

Where Are the Bones?

Imagine splitting a large slab at a fossil site, seeing a well-preserved vertebrate trackway inside, and also seeing at the end of that trackway a perfectly preserved skeleton (or body cast) of the actual animal that made the trackway: that is, imagine finding an animal literally "dead in its tracks." This might cause heart palpitations in a typical trackway collector, because there you would have the most direct evidence possible of what animal made that trackway. You would have a very good idea of what the animal looked like, how large it was, how many bones it had in its feet, how the foot structure connected directly to the trackway, and what kind of tail led to its tail trace, if present. But nature, unfortunately, is not that convenient. We are unaware of any reports documenting fossil skeletal remains from vertebrates together with tracks that the animal made just before it died, although such associations have been seen with invertebrates (Lockley and Hunt 1995).[6] This is because the conditions required for trackway preservation are not necessarily amenable to bone preservation, and vice versa. Also, fossil skeletons from vertebrate animals are orders of magnitude less common than invertebrate fossils. Vertebrates make up a minuscule portion of the animal kingdom, in both the number of species and the number of individuals.

Coal mines are especially poor places to look for body fossils. Since coal is made of compressed plant matter, plants would have been abundant in the area that became the mine. Areas rich in plants tend to have acidic soil (from plant decay), a condition that can easily dissolve fragile bones (and seashells). Insect body structures such as wings can be preserved in such an environment because the chitin that makes up insect exoskeletons, like the cellulose of plants, does not dissolve in acid.

Several fossil insect wings have been found at the Minkin Site (chapters 8 and 16). However, body fossils of terrestrial arthropods and of horseshoe crabs are delicate and rare.

If the bones of the Minkin Site tracemakers were not preserved where the tracks were made, then where should we look for them? According to Jun Ebersole, curator of Paleontology at the McWane Science center in Birmingham, Alabama, bones have been found at the Minkin Site, but only as small flecks in a marine layer that contains probable scallop shells (J. Ebersole, pers. comm.). These bone fragments may represent the remains of terrestrial animals that died and were washed to sea, buried quickly, and then fossilized. Marine sediment would be much less acidic than a coal swamp. The marine rock layer is one of only two kinds of substrate at a coal mine where bones might be found. The second place is in siderite nodules, brown concretions made largely of iron carbonate ($FeCO_3$) that form around organic matter. Some siderite nodules can contain well-preserved fossils, including bones. Siderite nodules from a coal mine near Carbon Hill, about 40 km (25 mi) from the Minkin Site, have yielded a well-preserved fish scale and other small fragments, and some of these nodules from Mazon Creek, Illinois, contain whole skeletons. Siderite nodules at the UCM have not yet yielded bone (Jun Ebersole, pers. comm.).

It is clear that at any paleontological site, we find only the fossils that the environment allowed to be preserved. This is called "preservation bias." Indeed, we can never really know the totality of organisms that lived in an area because of this bias. At a body-fossil site, animals that had only soft parts may not have been preserved at all. At a trace-fossil site like the Minkin Site, the only animals we find traces of are those that walked out of the nearby forest onto the muddy area to look for food. If there were animals that stayed away from the nearby mudflat and spent all their time in the forest, we would not find traces of these animals no matter how abundant they were. For example, the trace-fossil record in Alabama has not yielded any specimens of traces made by roaches, even though these would have been common in any coal-age forest.

Lockley and Meyer (2000) point out that even at sites where bones and tracks are found, the tracks were not necessarily made by the same creatures that left the bones. Again, this is a preservation bias issue.

Lockley and Meyer argue that tracks and bones preserve different representations of the fossil record, the former favoring terrestrial animals and the latter favoring aquatic animals.

Clearly, the "bone record" has strong limitations on what it can tell us about ancient life. It is indeed rare that the two domains, body fossils and trace fossils, ever overlap enough to give a consistent account of the fauna. One such rare exception is provided by the Tambach Formation in central Germany, where Lower Permian tetrapod skeletons and tracks are so well-preserved that trackmakers can be matched to tracks at the species level (Voigt, Berman, and Henrici 2007). *Orobates pabsti*, shown in fig. 3.1, top (Berman et al. 2004), is one of these matched skeletons. Interestingly, *body fossil casts* (as opposed to bone fossils) of 330-million-year-old amphibians were recently discovered in the collection of a Pennsylvania museum, in material donated before World War II.[7]

What's in a Name?

The naming of living organisms follows the well-known *binomial* rule or convention as defined by Swedish biologist Carolus Linnaeus in the eighteenth century. An animal or plant is usually assigned to a genus, or group of related organisms that share common characteristics, including morphology. Distinctive characteristics define species within the genus. For example, house cats belong to the genus *Felis* and the species *catus*, so the binomial name is *Felis catus*. The genus is always capitalized, while the species is always lowercase, and both are in italics (except when the name appears within italicized text, in which case it should be in regular type). The names are often Latin in origin and usually focus on some distinctive characteristic. For example, the genus name *Brontosaurus* means "thunder lizard."

Body fossils are named more or less the same way living organisms are named. One kind of skeleton will generally have one binomial name. Fossil plants are named somewhat differently, although ideally the same principle of one name per species applies. But fossil plants are usually fragmentary, and it may not be obvious that different fragments belong to the same genus and species. You may find a flower, a leaf, a root, a stem, or a bark impression. At first, you do not know which part goes with which other part. (You may never know.) So formal names for fossil plants often refer only to particular parts of plants. Later, scientists may discover which leaf, stem, flower, or other "bit" actually came from the same plant. We do not change the names, but we are mindful that when we find a leaf of one plant we should expect to find its stems and flowers as well. As an example of how complicated plant fossil names can be, *Lepidodendron* is a common genus of giant arborescent lycopods (scale trees). It is a distinctive type of bark impression (chapter 17). Also commonly found are root casts called *Stigmaria* and upper branch fossils called *Lepidostrobophyllum*. These were named when found separately, but all are parts of the same species or genus of tree.

The naming of trace fossils also follows the binomial convention but uses the terms "ichnogenus" (capitalized) and "ichnospecies" (lowercase), where "ichno" is short for "ichnological" (meaning trace fossil). Like plant and animal body fossils, trace-fossil names refer to a collection of specimens that all look alike in particular ways. For example, the ichnogenus *Undichna* refers to sinusoidal grooves made in sediment. We infer that they were all made by the fins of fish, but we usually cannot say which genus and species of fish was responsible for a given occurrence of *Undichna*. Even when fish with quite different skeletons make grooves in the mud with their fins, the grooves may not look very different.

An example with a full binomial name is *Cincosaurus cobbi*, which could be translated as "Cobb's five-toed lizard" (chapter 7). This name refers to a type of fossil vertebrate trackway that was discovered in Walker County. It is not the genus and species of the animal that made the tracks, because more than one type of animal can make similar-looking tracks. It merely implies that a reptilian tetrapod having five toes on both hind feet and forefeet made the tracks. In ichnological terms, *C. cobbi* is an ichnotaxon, one of five tetrapod ichnotaxa specific to the site (chapters 14 and 15).

Fossil species frequently get renamed, for two main reasons. The first is a matter of bookkeeping. Sometimes, two different people give the same fossil different names. This happens because the second scientist did not know about the name already published by the first scientist. Perhaps the first publication appeared long ago, in an obscure journal, or in a language the second scientist could not read. It is also possible for the same scientist to assign two names to the same animal based on different sets of fossils. Whatever the reason, we use only the first assigned name, which we say has scientific priority.

A classic example of one type of organism getting two names is the familiar and very popular dinosaur called *Brontosaurus*. These large sauropods were in the limelight during the early '60s when the popular animated series *The Flintstones* was on evening television. Although *Brontosaurus* was recognized in 1879 by O. C. Marsh, later studies of *Brontosaurus* bones and the bones of another animal, *Apatosaurus*, which Marsh described in 1877, suggested that they were the same animal.[8] Thus the name *Brontosaurus*, although popular, is technically incorrect because the first-coined name, *Apatosaurus*, always takes priority.

The other reason a fossil species gets renamed is to correct a mistake, which is often the result of misinterpretations caused by poor specimen preservation, or by a very limited number of samples. Such mistakes can be corrected when better or more specimens are found. In the case of trace fossils, for example, the same kind of trackway can be preserved or exposed in different ways and get multiple binomial names (called synonyms). If a species has been assigned to the wrong genus, it will be transferred to an existing genus, or assigned to a new genus, if no existing genus has sufficiently similar morphology.

Note that the naming of genera and species (or ichnogenera and ichnospecies) is a specialized field, and a delicate matter that can require considerable research. Determining that a given body or trace fossil is new to paleontology generally requires an extensive literature search, sometimes involving visits to museums or other places that house reference specimens of known species. Most species are represented by holotypes. A holotype is a single specimen formally designated as the "best" or most characteristic representative of a given species.

We have examined how trace fossils and body fossils form and get named, and how the two domains of study rarely overlap to any great degree. Let us now examine the UCM fossil assemblage in terms of the evolution of reptiles and amphibians, and the geological history of Alabama.

4

Amphibians to Reptiles

Paleozoic sedimentary rocks in Alabama provide us with extraordinarily detailed insight into the behavior of wildlife that lived in the state during the coal age. Nature has provided no other way for us to view or even know about this wildlife, because no bones of these animals have yet been found. All we have is the traces to tell us that once upon a time, Alabama was a tropical paradise teeming with primitive four-limbed animals. The thrill of finding the traces that these extremely ancient animals made when they were alive and going about their daily activities is surely a driving force for many avid ichnologists, whether amateur or professional. But in the case of the Minkin Site, there is an even greater reason to find the tracks interesting: the wonderful world of dinosaurs that we all know and love, that is, the world of reptiles, got its start with animals like those that lived at the Minkin Site. The Minkin Site animals lived at precisely the time when reptiles were emerging as dominant vertebrate land animals.

To appreciate this, we first need to know a little about how ages of fossils are determined. As any paleontologist will attest, the age of a fossil is very important, because without it we cannot piece together a reliable picture of the history of our planet. When we know the age, we gain a perspective that allows us to draw in other information, such as the characteristics of fossils of the same age that were found in other places. We can ask questions about ecosystems that may have been prevalent at the time and that may explain something about a fossil's characteristics. We can tie a given fossil to important events in Earth's history and see how it fits into the evolution of life on Earth. Age opens a door to a vaster array of knowledge for any given specimen.

Even someone only casually acquainted with paleontology knows that fossils can be extremely old compared to any human time scale. For example, people who attend the Alabama Museum of Natural History's annual summer field trips to Shark Tooth Creek, near Tuscaloosa, are told that the shark teeth are 80 million years old. The dark and often still very sharp teeth date back to the Cretaceous Period, the last period of the Age of Dinosaurs. For the young people involved in these field trips, hearing that the tooth they hold in their hands is from a shark that lived 80 million years ago will usually elicit a response such as "awesome!" It is difficult, if not impossible, to really understand such a great age in human terms. It is four million human generations, more than ten thousand times as old as human civilization, and much, much older than dirt!

In chapter 2, we noted that the *Cincosaurus* beds (the layers of rock containing vertebrate trackways like *Cincosaurus cobbi*) at the Minkin Site are Westphalian A in age, about 313 million years old (Pashin 2005). How do we know this? There are accurate and precise radiometric dates (that is, ages based on relative amounts of radioactive elements) of Westphalian strata in other parts of the world, but not from Walker County. These well-dated rocks in other places contain *index fossils*, the remains of organisms that existed for only brief periods (typically only hundreds of thousands of years). They are diagnostic of rocks laid down during the particular time when they lived. In the case of the Carboniferous, many index fossils are plants. Some of these plant fossils are found at the Minkin Site, linking the coal-bearing strata there to well-dated rocks elsewhere. Index plant fossils are found on the same pieces of rock as fossil trackways at the Minkin Site.

The best coal-age index fossils are not necessarily easy to see. They are spores and pollen, which require compound (high magnification) microscopes to be seen. The arborescent lycopods (the extinct giant "scale trees" known for the beautiful patterns of leaf scars on the surfaces of their trunks and stems; chapter 17) reproduced by spores, while pteridosperms (seed ferns, extinct nonflowering plants only distantly related to modern ferns) used pollen and produced seeds, structures adapted for reproduction on dry land. Fossils of the pteridosperm pollen organ *Whittleseya elegans* are common at the Minkin Site.

Of all the parts of the Pottsville Formation, the Mary Lee coal zone (which includes the Minkin Site tracks; chapter 9) is the best dated. Geologist Richard Winston (1991) used coal balls (roughly spherical clumps of petrified plant matter found in coal seams) from the Mary Lee coal zone to identify well-preserved plant fossils from the Westphalian A.

"Extinct" is a popular and familiar term in paleontology. In addition to knowing the age of a fossil, most people also want to know whether the organism it represents is extinct. The answer is usually yes; something like 99 percent of all species that have ever lived have gone extinct. Fossil sites do not, in general, include fossils of living species unless the rocks exposed at the sites are very young.

TETRAPODS AND THE HISTORY OF LIFE

Tetrapod tracks at the Minkin Site come in two varieties: those made by amphibians and those made by amniotes, the latter referring to animals with a more advanced mode of reproduction. Amphibians can live on land but generally lay their shell-less eggs in water. Amniotes are terrestrial tetrapods that lay their eggs on land. Early reptiles get a lot of attention from researchers because they were the *first amniotes*. Today, amniotes include not only reptiles but also mammals and birds.

We know that both amphibians and reptiles lived at the Minkin Site because of the number of digits in the forefeet of the fossil tracks. Most amphibians, even those from the Carboniferous Period, have only four toes on their front feet, or manus limbs. In contrast, reptiles have five manus toes.[1] Both types of animal have five toes on the hindfoot (pes limb). Tetrapod trackways collected from the Minkin Site include both four-toed and five-toed manus impressions (Haubold et al. 2005a); Figure 4.1 shows two examples. The left panel shows a type of amphibian trackway from the Minkin Site called *Nanopus reidiae*, while the right panel shows a selection of *Cincosaurus cobbi* tracks from the Aldrich and Jones type specimen (Aldrich and Jones 1930, plate 7). All the amphibian trackways recognized at the Minkin Site were made by creatures with four toes on each manus, while the animals that made the *C. cobbi* tracks had five clear manus digits and could not have been amphibians. There are also other differences. Amphibians have a wider, more sprawling gait than reptiles, and amphibians also do not have claws, whereas

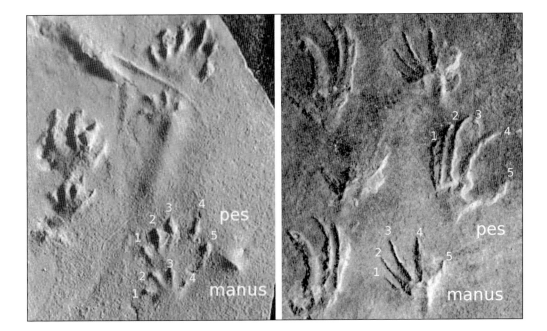

reptiles do. In some localities reptile claw marks are preserved, but we do not have definite examples of claw marks from the Minkin Site (largely because claw marks cannot usually be recognized in the undertracks of small animals). Finally, the forefeet of amphibians are often noticeably smaller than their hind feet; this is not necessarily true of reptiles.

Knowledge of the age of the Minkin Site fossils helps us put them into their evolutionary context. The transition of vertebrate animal life from water to land is believed to have begun at least 380 million years ago, during the Devonian Period, commonly known as the Age of Fishes. This great evolutionary leap made by some fish, from locomotion by *swimming* in water to locomotion by *walking* on dry land, is one of the truly remarkable innovations of vertebrate life and is beautifully described in the book *Gaining Ground: The Origin and Evolution of Tetrapods*, by Jennifer A. Clack (2012). At the beginning of the Devonian Period, 416 million years ago, plant life was restricted mostly to shore areas. There was no grass, and no insect noise on land. The land was eerily silent and, apart from perhaps lichen and mosses, mostly barren.

The key to tetrapod development was the bone structure in the fins

FIGURE 4.1 Comparison between reptile and amphibian tracks. The left frame shows UCM 1141, the holotype (defining specimen) of *Nanopus reidiae*, the trackway of a small amphibian. The right frame shows the holotype of *Cincosaurus cobbi* (chapter 7). The distinction is highlighted by the numbering of the digits in the manus and the pes. *N. reidiae* has four digits on its manus print, while *C. cobbi* has five. Often, in *C. cobbi*, only digits 2, 3, and 4 of the pes are visible. Both specimens are positive hyporelief.

of certain kinds of fish that also had the ability to breathe air. Most fish, including modern ones, have left-right pairs of fore (pectoral) and hind (pelvic) fins. They also have dorsal (upper back), anal (lower rear), and caudal (tail) fins, which mark the body's midline and do not come in left-right pairs. Certain Devonian fish called *lobe-fins* had pectoral and pelvic fins that were attached to their shoulder and pelvic girdles by a *single radial bone* in an arm- or leg-like manner. Other fish (called ray-fins) had multiple, roughly parallel radial bones connecting the fins to their girdles (Clack 2012).

The fins of lobe-fins, with their sturdy connection to the torso, could act like limbs, pushing hard against the sea, estuary, or river bottom. Lobe-finned fish could likely crawl along the bottom of the sea and bury themselves in mud, but in addition, they could raise their heads above the water near shore. They could drag themselves right out of the water and look around. Perhaps environmental pressures, such as predatory fish, the drying of isolated pools in which they became trapped, the lure of warm shallow water, or the abundant arthropods (invertebrate life) near the lush, vegetated freshwater shorelines, caused some descendants of these fish to leave the water permanently and live on land. By the time this happened, plants and invertebrates had occupied the land for millions of years. Clack argues that lobe-finned fish were emboldened to venture onto land only after the plant canopy became thick enough to protect them from the hot sun, provide a humid environment, and support a significant supply of invertebrates for food. Lobe-finned fish represent only one model for fin growth, but ultimately it was not very successful (most modern fish are ray-fins). However, lobe-fins turned out to be preadapted for terrestrial life. Both lobe-fins and ray-fins used pectoral and pelvic fins for locomotion in water, but only the lobe-fins had the bone structure that allowed these same fins to successfully adapt to locomotion on land.

The transition of animal life from water to land was a momentous event in the history of life on Earth. Tetrapods eventually came to dominate the land, at least as far as large animals are concerned. According to Clack, the unique attributes that define tetrapods as a natural group point to a single lobe-finned fish as the *common ancestor of all tetrapods* and of any animals whose ancestors were once tetrapods. Thus, human

Figure 4.2 Artist's conception of *Tiktaalik*, a transitional animal between fish and tetrapods.

beings are tetrapods by descent even though, at least after we exit toddlerhood, we are generally bipedal. Snakes are tetrapods also; the common ancestor they share with lizards had four legs. A transitional life form between the lobe-finned fishes and tetrapods is the remarkable fossil *Tiktaalik*, beautifully described in Neil Shubin's 2009 book *Your Inner Fish: A Journey into the 3.5-Billion-Year History of the Human Body*. *Tiktaalik* looked more like a tetrapod than a fish (fig. 4.2), but it was still a fish. This led Shubin to refer to it as a "fishapod" because it had so many characteristics in common with both fish and tetrapods. This discovery was a boon to the understanding of the evolution of tetrapods, and *Tiktaalik* is one of the most important transitional fossils ever found.

The Westphalian A was almost 60 million years after early fish left their lakes and eventually gave rise to amphibians. Amphibians had already been around a very long time when the Minkin Site tetrapods were alive, almost as long as the time, from our perspective, since dinosaurs died out. But not so for amniotes, according to Clack: "It is during the Westphalian that the first remains of animals that can be confidently called reptiles are found." Since this matches perfectly with the time frame of the Minkin Site rocks (as well as with those in Carbon Hill and throughout Walker County), we can confidently assert that *the Minkin Site trackways represent some of the earliest evidence for reptiles in the world.* This is what really makes the Minkin Site fossils special and important. The tetrapod trackways from the Minkin Site include some of the oldest reptilian footprints yet known (Falcon-Lang, Benton, and Stimson 2007; Falcon-Lang et al. 2010; Keighley et al. 2008).

Reptilian and Amphibian Radiations

The occurrence of both reptilian and amphibian trace fossils at the Minkin Site is significant. The earliest reptiles were breaking new ground: they were occupying new ecological space. Their eggs probably had hard shells. Reptiles could survive and grow in dry places where amphibians never could, just as seeds permitted some plants to move into environments that had been barren for billions of years. This meant that reptiles could move into areas that had previously not been inhabited by vertebrates at all.[2]

What did the earliest reptiles do? Where did they go? How did they interact with amphibians in places where the two coexisted? It might be possible to answer some of these questions by detailed analysis of the trace fossils at the Minkin Site, which is one of the oldest sites at which reptiles and amphibians are known to have lived together. Two other mine sites in Walker County, the former Kansas Mine and the Sugar Town Mine, have also yielded both reptile and amphibian trackways.[3]

According to Clack (2012), diversification among tetrapods began during the Early Carboniferous Period and intensified during the Late Carboniferous. Westphalian localities in Europe and North America, including the Minkin Site, have provided evidence that tetrapods moved into freshwater, and that coal swamps affected by tidal conditions were perfectly suited to the creatures that made the tracks. Figure 4.3 shows skeletal reconstructions of the three main types of vertebrate animals that left footprints at the Minkin Site: temnospondyl amphibians, anthracosaurs, and amniotes. These reconstructions are based on body fossils collected from other parts of the world (Clack 2012) but

Balanerpeton - temnospondyl amphibian

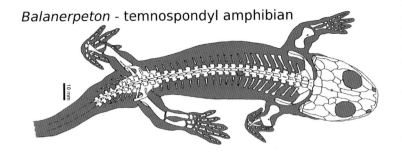

10 mm

Silvanerpeton - anthracosaur

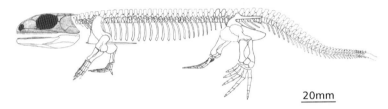

20mm

Paleothyris - amniote

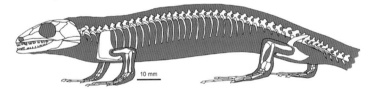

10 mm

FIGURE 4.3 Skeletal reconstructions of three kinds of tetrapods based on bones collected from other parts of the world. *Top*, artist's conception of *Balanerpeton*, an Early Carboniferous temnospondyl amphibian. *Middle*, schematic of *Silvanerpeton*, an anthracosaur. Anthracosaurs are thought to be closely related to both amniotes and amphibians. *Bottom*, schematic of *Paleothyris*, an early amniote with five toes on the manus.

are probably good representatives of the animals that lived in this part of Alabama during this time.

During the Westphalian, temnospondyls became the dominant group of amphibians. These amphibians had a large opening in the palate (base) of their skull, a characteristic that unites them with modern salamanders and frogs. This feature may have enabled the creatures to better swallow their food or breathe air (Clack 2012). The animals had large eye sockets and their hind limbs were longer than their forelimbs, as in modern amphibians. There was also significant variation in skull morphology. All the amphibian tracks from the Minkin Site were probably made by temnospondyls, according to Haubold et al. (2005a). Minkin Site trackways attributed to amphibians are *Nanopus reidiae* and *Matthewichnus caudifer* (chapter 15). *Balanerpeton* was a temnospondyl of Early Carboniferous age that might have resembled Minkin Site trackmakers (fig. 4.3, top). The skull was 2.5 to 5 cm long, and the body ranged from 18 to 44 cm in length. *Balanerpeton* body fossils were abundant at a well-known fossil site near Edinburgh, Scotland. Clack (2012) indicated that this site contained the earliest known terrestrial assemblage of tetrapod species.

Anthracosaurs are also represented in the Minkin Site ichnofauna. These somewhat mysterious animals, called "coal lizards," formed a transitional group that showed characteristics of both temnospondyl amphibians and amniotes. *Silvanerpeton* (fig. 4.3, middle) was an Early Carboniferous anthracosaur found at the same locale as *Balanerpeton*. *Silvanerpeton* was a small animal, but Clack (2002) shows a reconstruction of a Late Carboniferous anthracosaur, *Pholiderpeton*, that was up to 4 m long. In chapter 15 we describe the tracks of *Attenosaurus subulensis*, thought to have been made by an anthracosaur as much as 1.5 m in length. The smaller trackways, *Notalacerta missouriensis* and *Cincosaurus cobbi*, found at the Minkin Site have been attributed to amniotes by Haubold et al. (2005a). *Paleothyris* (fig. 4.3, bottom; Clack 2012) was an amniote comparable in size to the animals that made these trackways. Both amniotes and anthracosaurs had five digits in the manus limbs.

The terrestrial vertebrate fauna at the Minkin Site was thus quite diverse, containing representatives of three major groups. The anthracosaurs were by far the largest animals and were probably top predators. The abundance of amphibians, anthracosaurs, and amniotes at the Minkin Site probably reflects the location of the site at a time when all three groups of organisms were important in the world, and at a place where land, sea, and freshwater (rivers) met. Such an environmentally complex setting encouraged development of a complex biota. Interestingly, other sites in Walker County of about the same age seem to be less diverse, such as the old Galloway Coal Company No. 11 Mine (chapter 7) and the recently active Crescent Valley Mine (chapter 19) sites near

Carbon Hill. These cases may be attributable to subtle environmental differences, such as water salinity (Buta et al. 2013).

Landscapes in Alabama

The backdrop of the evolution of tetrapods worldwide is the history of the land, because that is where most tetrapods live. How much we know about that history depends on how it has been recorded through time. However, the land is not a good place to record information. The geologic history of the Earth's surface is written in sedimentary rocks (Lacefield 2000), but most terrestrial environments are erosional. How much sediment is permanently stored on a hillside, or on a city street? The formation of sedimentary rocks depends on the deposition and accumulation of sediment. To accumulate sediment, and to preserve the geologic record, you need a low and quiet place—like the bottom of a sea, or a coal swamp.

Three forces do most of the work of moving sediment: water, wind, and gravity. Of these, water is by far the most powerful. Frozen water in the form of glaciers has great capacity to move sediment, but liquid water is most prevalent on Earth. Because liquid water moves downhill, sediment carried by water tends to end up on the ocean floor. Runoff from a rainstorm picks up sand or mud from your yard or driveway and washes it into the closest creek. From there it is carried to a larger river, and thence to an even larger one, always heading downstream. Sediment can be stored in river point bars and floodplains, in lake beds, in bays, and in swamps. Estuaries form where some rivers meet the ocean, and estuarine deposits, like those in Mobile Bay, are common in the rock record. In this book, we argue that the *Cincosaurus* beds at the Minkin Site formed in an estuarine environment.

We know that sedimentary rocks deposited during Westphalian A time are exposed in Walker County, and that these rocks record the footprints of some of the earliest reptiles. But how do these rocks connect to those found elsewhere in Alabama? Why do we not find the same kinds of rocks near Mobile, Gadsden, or Demopolis?

To answer these questions, we need to examine a *geologic map* of Alabama. This type of map shows how the rocks exposed in different parts of Alabama represent different periods in Earth's history. Figure

4.4 shows that the geology of Alabama is complicated. The state can be divided into distinct regions based on the age and characteristics of the rocks. The triangular area in eastern Alabama labeled "P" is the Piedmont. It contains the oldest rocks in Alabama, dating back more than 500 million years. This is the southwestern tip of the core of the Appalachian Mountains, which stretches all the way to Canada. Piedmont rocks are highly altered after having been deeply buried, squeezed and heated, and then pushed back to the surface by the collision of North America with the African portion of Gondwana during the Carboniferous Period (chapter 5). The rocks have been fractured and folded and contain mineral crystals that grew in place far beneath the surface of the Earth. Although most Piedmont rocks began as typical sedimentary rocks, geologic events and processes changed them into what geologists call metamorphic rocks. Intense heat and pressure transformed them profoundly but did not completely obscure their origins.

Northwest of the Piedmont is a strip of Alabama labeled "VR" on figure 4.4. This stands for Valley and Ridge, a descriptive term for a region composed of long, straight mountains separated by long, straight valleys. These rocks are also part of the Appalachian Mountain belt, but they have not been so drastically transformed as metamorphic rocks have, only folded, faulted, and fractured. Some of the most interesting rocky outcrops you will see consist of formerly flat sandstone or limestone that has been pushed from the southeast and wrinkled like a rug.

Farther west ("AP" and "ILP") are ancient plateaus made of layered rock that was far enough from the locus of continental collision to remain horizontal. Nearly all the rocks north of an arc extending from the vicinity of Auburn to the northwestern corner of Alabama are old, 300 million years or more, and most were laid down in or adjacent to ancient seas. In these seas were armored fish like none found anywhere today, ammonites (squid-like animals in spiral-shaped shells that could be up to 2 m [6 to 7 ft] in diameter), and many other creatures that went extinct with all their kin before any dinosaur walked the Earth. What did not yet exist anywhere were mammals, birds, or any of the kinds of fish we find today. Exposed land harbored strange forests full of plants only distantly related to modern vegetation.

The greater part of Alabama is underlain by part of the Gulf Coastal

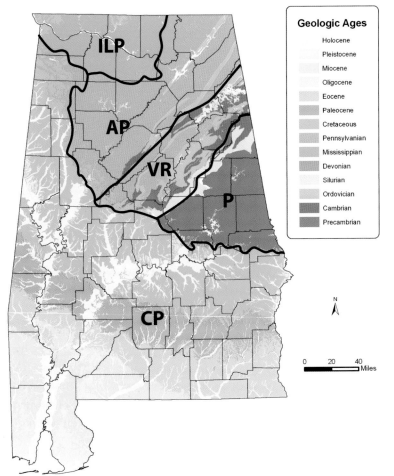

FIGURE 4.4 Geologic and physiographic regions of Alabama.

P = Piedmont
VR = Valley and Ridge
AP = Appalachian
        Plateau
ILP = Interior Lowland
        Plateaus
CP = Coastal Plain

Plain ("CP" on the map), a wedge of sedimentary rock that extends from South Florida all the way around the Gulf Coast and into Mexico. The Coastal Plain consists of all the sediment that has been washed off the continent into the Gulf since the time of uplift of the Appalachian Mountains, plus everything that formed from the minerals in seawater in the peripheral part of the Gulf itself (seashells, reefs, rock salt), except for the part that is still underwater out in the Gulf. This has been going on since the Gulf was born, more than 200 million years ago. Coastal Plain sediment is thick, and it gets thicker the farther south you go. Sediment deposited on the Coastal Plain ranges from a very thin area south of Birmingham to more than 9.5 km thick in the vicinity of Mobile. On the continental shelf south of Alabama the submerged extension of the Coastal Plain is nearly 13 km thick in places. As you move toward the Gulf, the surface geology gets younger and you find fossils of more recent types of animals (for example, shells that look like those on the Gulf beaches). The sediment surface, from the latitude of Tuscaloosa all the way to the edge of the continental shelf, slopes downward from about a hundred meters (several hundred feet) above sea level to about a hundred meters (several hundred feet) below. The northern edge is a sort of hinge, and everything farther south sinks under its own weight. The parts far from the hinge sink faster, which leaves more room for new sediment on top. This just makes the thick end of the wedge get thicker, and sink faster, which is why younger material is preserved in the south, away from the hinge.

Where does the Minkin Site fit into this complex geology? South of the ancient plateaus and west of the mountains lies a small area called the Black Warrior Basin. This area sank while the mountains were rising to the east. The basin filled with a wide variety of sedimentary rocks that included the remains of whole forests and produced a great deal of coal. In the same way, the Gulf of Mexico has been filling with sediment brought in by the Mississippi River and other rivers. Sedimentary rock in the Black Warrior Basin approaches two kilometers in thickness, attesting to the vast volumes of sand, mud, and other material that poured into the basin from the Appalachian Mountains to the east (Pashin 2005).

We have highlighted in this chapter how Minkin Site trace fossils in particular, and Walker County fossils in general, are directly connected to the earliest known reptiles. From these primitive forebears came all the land animals that followed, including dinosaurs, mammals, and birds. When we behold the ancient footprints of Alabama, we are looking through the lens of deep time to an event that is directly relevant to our own existence.

We have also shown that only in a small part of Alabama are the rocks that contain the ancient footprints exposed. If these rocks had eroded a million years earlier, we might not know anything at all about the animals that lived in Walker County. The rocks in northeastern and northwestern Alabama are too old, and those in southern Alabama are too young, to tell us anything about them. Now let us look further into the world of the Minkin Site trackmakers.

5

The Pennsylvanian Period in Alabama:
Looking Down

Paleontologists and geologists generally regard the Pennsylvanian Period as one of the best-recorded periods in Earth's history. This is because the rocks and fossils from the coal-age swamps that dominated the tropical landscapes are so abundant that we know a great deal about the life and the environment at that time. It is not surprising that land-dwelling vertebrates (reptiles) evolved rapidly during this time, because the coal-age forests richly supported them with the necessities of life. The forests provided shade, places to live, water, and abundant invertebrate life for food. Since the Pennsylvanian Period provides the setting of our story, here and in the next chapter we want to take you on a journey back in time. What was the world really like back then, and what might we see if we visited?

The Pennsylvanian Period happened because geologic forces made it happen. In our solar system, the Earth is the largest *terrestrial planet*, a type of planet made mostly of rock and metal, orbiting relatively close to the sun. This is in contrast to giant planets like Jupiter and Saturn, which are made mostly of light gases and liquids and orbit very far from the sun. All planets generate their own internal heat, and their ability to keep this heat inside for a long time depends on their mass. In the case of the Earth, at least half of the internal heat is produced by the decay of radioactive elements that emit subatomic particles or energetic photons (particles of light) into the surrounding rock.[1] These particles inject energy into the rock, which over a long period can produce internal melting and contribute to what is known as the "differentiation" (division into layers of different density) of the Earth's interior. As the most massive terrestrial planet, the Earth has the largest amount of internal heat of all the terrestrial planets and also the greatest ability to hold that heat inside for a long time.

The Earth cannot, however, keep its internal heat bottled up indefinitely, because heat naturally dissipates (flows in the direction of hot to cold). If the inside of the Earth is hotter than the surface, then the internal heat must flow from the inside to the surface. The main way the heat flows is by *convection*, which refers to heat transport by the mass movement of hot material. Generally, heat is transported by convection when rising bubbles of hot, lower-density material reach the surface, release heat there, cool, and become dense enough to sink. Convection is a cyclic process, meaning the rising and sinking occur over a period of time. In the case of the Earth's interior, the process is very slow, on the order of a hundred million years per cycle.[2]

Differentiation led to the Earth having a crust of lower-density rocks that lies on a partially molten layer of rock below. The crust consists of both continental (land) and oceanic (sea-bottom) rocks, which together with upper mantle rocks are part of what is called the *lithosphere*. It was realized in the mid-twentieth century that the Earth's lithosphere is broken into a small number of pieces called *plates*, and that these plates can actually move around because of the plastic nature of the underlying rock layer. A rising convective cell of hot material from inside the Earth can produce an oceanic ridge, an elevated sea-bottom zone where two plates are moving apart. As the plates separate, new rock forms on either side of their boundary. The outward movement of rock from a ridge creates pressure elsewhere. New rock comes to the surface at ridges (divergent plate boundaries), while at convergent boundaries the edges of some plates are pushed down (subducted) under other plates. Gravity and convection both contribute to draw subducted rock down inside the Earth.[3] Geologists call this continental movement *plate tectonics*, and it is now a well-known process. It is slow but measurable, and it explains large-scale events that took place during the Pennsylvanian Period in Alabama.[4]

When the Minkin Site trackmakers were alive, most of Earth's

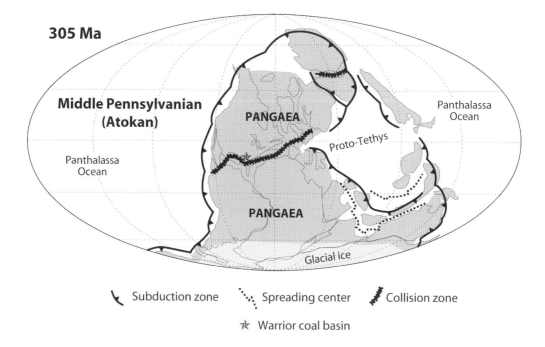

305 Ma

Middle Pennsylvanian (Atokan)

PANGAEA

Panthalassa Ocean

Proto-Tethys

Panthalassa Ocean

PANGAEA

Glacial ice

⎣ Subduction zone ⋯⋰ Spreading center ✦ Collision zone

✫ Warrior coal basin

FIGURE 5.1 Schematic showing the supercontinent Pangaea. Different kinds of plate boundaries are labeled. A subduction zone is where one plate is sliding under another, while a spreading center is where two plates are moving apart. A collision zone is where mountains are building. The Warrior coal basin is where the coal reserves of Alabama are located. This basin (indicated by the large star) was south of the equator at the time. The Atokan is a North American stage comparable to the Westphalian in figure 2.1*A*.

continents had gathered into a supercontinent called Pangaea. There was no Atlantic Ocean, because Europe and North America were in contact; you could walk from New England to old England. The combined European and North American section of Pangaea, called Laurussia (or Euramerica), had collided with the southern section, called Gondwana (fig. 5.1).[5] As the continents came together, huge masses of rock were shoved across the landscape, their weight depressing the land they covered and everything nearby as well. This is how the Black Warrior Basin was formed.

Paleontologic and geologic studies allow us to trace the locations of ancient shorelines and other geologic features, a field known as paleogeography. To do this reliably requires knowledge of how the Earth's magnetic field interacts with rocks, how fossil species are distributed around the world, what rocks and fossils imply about ancient climates, and how plate tectonics shapes the Earth's surface.[6]

Figure 5.2 is an artist's rendition of the paleogeography of the Black Warrior Basin in the Mississippi and Alabama region.[7] The present-day

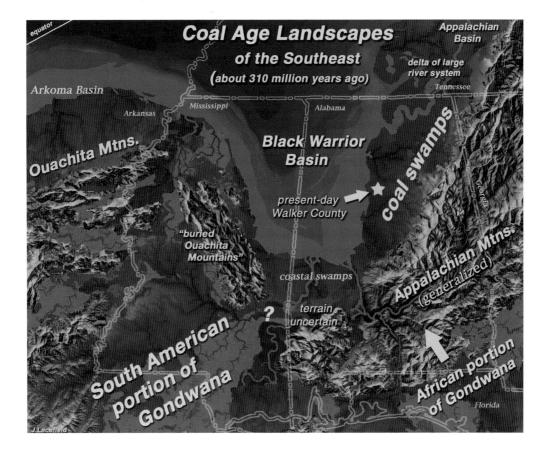

Coal Age Landscapes
of the Southeast
(about 310 million years ago)

Appalachian Basin

delta of large river system

Tennessee

Arkoma Basin

Arkansas

Mississippi

Alabama

Ouachita Mtns.

Black Warrior Basin

present-day Walker County

coal swamps

"buried Ouachita Mountains"

coastal swamps

Appalachian Mtns. (generalized)

terrain uncertain

?

South American portion of Gondwana

African portion of Gondwana

Florida

J.Lacefield

outlines of these states and parts of Louisiana, Georgia, and Tennessee are indicated. The basin is named after the Black Warrior River system that currently drains it. The map shows the basin when it was filled with water and how the huge tropical swamps were laid out around it. Note the location of the equator at the upper left. Alabama was slightly south of the equator during the Pennsylvanian, prior to its slow drift northward to its present location. The Black Warrior Basin was a product of the collision of Gondwana with Laurussia, forming in a "pocket" where the growing Appalachian and Ouachita Mountains met. As the land surface in the basin sank, water did what it always does: it sought the lowest place, bringing sand and mud again and again to build up cyclic deposits that we call the Pottsville Formation. In addition to providing Alabama

FIGURE 5.2 Paleogeographic reconstruction of the southeastern United States during the coal age. Walker County, Alabama, where the Minkin Site is located, is indicated by a star.

with coal, the ancient swamps produced methane (natural gas) and preserved a detailed record (through sedimentary rocks) of the animals and plants that once lived where the land met the ancient inland sea.

The Pottsville Formation, which comprises all coal-age rocks in Alabama, records the weathering and erosion of the rapidly uplifting mountains, the transport of sediment by rivers through deltas into adjacent ocean basins, and the formation of vast swamps, estuaries, and tidal flats in the coastal regions right in the midst of it all. The rocks that formed in this area constitute the coal measures of the Black Warrior Basin.

At the time Laurussia and Gondwana were colliding, the South Pole was covered by part of Gondwana. Because of periodic changes in the ellipticity of the Earth's orbit, as well as changes in the tilt of the Earth's rotational axis, this portion of the planet experienced a glaciation that advanced and retreated in much the same manner as the most recent glaciations only a few tens to hundreds of thousands of years ago. Cyclic glaciations caused global sea levels to rise and fall periodically. When glaciers advanced, sea levels dropped low enough to expose new swampy land in the tropics. This facilitated the growth of the huge tropical forests. The forests pumped so much oxygen into the air that the atmospheric oxygen level reached 35 percent, the highest it has ever been (Stanley 2009). As we noted in chapter 2, when plants died in these massive forests, they accumulated on the swamp bottom and transformed into what is known as peat, or partly decomposed plant matter. Peat forms when a high volume of plant material accumulates in shallow water, as it would have in a coal-age tropical swamp forest. Instead of rotting away completely, the plants simply lay in stagnant, oxygen-poor water until they were covered by layers and layers of mud. There was so much organic material in the swamp water that, in spite of the high amount of atmospheric oxygen, the water could not maintain enough dissolved oxygen to allow bacteria, fungi, and other organisms that normally turn plant debris into rich soil to fully break down the dead plants. As long as the forest stayed alive, the peat would merely accumulate in the swamp, piling up to a thickness of 3 to 6 m (10 to 20 ft) over time (Pashin 2005). But because of global climate cycles, the planet eventually warmed, the glaciers over the South Pole receded, and sea levels rose enough to flood the tropical forests. The flooding killed the forests and piled sediment on top of the peat. Enough sediment accumulated to compress the plant material and, over time, convert it into coal.

The lower Pennsylvanian Pottsville Formation is dominated by shale, sandstone, and coal. The upper part of the formation exceeds 1830 m (6,000 ft) in thickness (McCalley 1900; Butts 1926; Thomas 1988) in the Black Warrior Basin, and 2440 m (8,000 ft) in the Cahaba coalfield to the southeast (Pashin et al. 1995). These rocks were deposited in a river-delta setting and display excellent cyclicity of different rock types in a predictable sequence (Ferm, Ehrlich, and Neathery 1967; Gastaldo, Demko, and Liu 1993; Pashin 1998). In general, coal forms in seams (beds) that are typically about 0.5 to 1 m (1.5 to 3 ft) thick. Coal becomes the strange dark substance that it is not only because the plant material is compressed, but also because it is partially heated (cooked). A coal seam 1 m (3 ft) thick would once have been a peat layer 10 m (30 ft) thick, with the remains of millions upon millions of trees and smaller plants. As the dead plants were buried deeper and deeper in the Earth, the slow heating and tremendous weight of sediment on top of them caused chemical as well as physical changes. Volatile gases, that is, atoms of oxygen and nitrogen, among others, were gradually stripped away or "driven off" by the pressure. Carbon, the backbone of all organic molecules, and hydrogen were left behind and concentrated. Coal formation is a complex process and can result in different *grades* depending on the carbon content, but the result is usually a black substance made mostly of carbon with lesser amounts of hydrogen, sulfur, and heavy metals like mercury. The coal in the Black Warrior Basin has a banded appearance and is a medium-grade coal called bituminous (chapter 2).

The Mary Lee coal zone contains four major coal seams (chapter 9). Coal seams are almost always closely associated with clay layers that contain less plant material but generally better plant fossils; they represent the accumulated life, and death, of a tropical swamp forest. How do we explain multiple coal seams, as in the Mary Lee coal zone (fig. 2.2*A*)? The answer is also tied to South Pole glaciation. During the next glacial period, the sea would recede, a new forest would grow, and a new peat layer would form. Later, as the ice melted and the sea level rose again, this younger forest would also be killed and turned into another compressed seam of coal in the rock layers, but this time several meters *higher* than the previous seam. Coal would form in cycles as long as there were climatic oscillations, sufficient land over the South Pole to store glacial ice for long periods, and other factors that allowed the swamps to exist. Eventually, however, Pangaea broke up and the coal-age forests disappeared, and many of their most important plant species (like the seed ferns and arborescent lycopods) became extinct.

Another important aspect of life in Alabama during the coal age was the higher average temperature, high humidity, and lack of seasonal variation that resulted from the future state's location in the tropics at that time. Today there is other land in the tropics, and Alabama lies at a latitude where seasonal variation is significant. The plants that lived in coal-age Alabama could not thrive at our current latitude. The state is swampy only near its south coast, the inland sea is gone, and the plants we see now are those that can survive the seasonal variation we now experience. After the coal-age forests disappeared, the oxygen level in the Earth's atmosphere dropped significantly; it stands today at 21 percent.

Figure 5.3 shows the paleogeographic setting of the Minkin Paleozoic Footprint Site during the Pennsylvanian Period (Lacefield 2000). The site was near the shore of an estuary, a transitional region where a river meets a tidally influenced sea. In the case of the Minkin Site, paleogeographic analysis indicates that an ancient river flowed through the area from the nearby Appalachians into the Black Warrior Basin. The map also shows the broad extent of the coal swamps, and how the Appalachian Mountains were much higher south of Tuscaloosa than they are today anywhere in the state.

Pennsylvanian Life in Alabama

Now that we have some idea of what Alabama was like during the coal age, we can begin to imagine what we might see if we could go back to that time. If we were to fly over north-central Alabama (say from ancestral Atlanta toward ancestral Walker County), we would be impressed first with a mountain range as high as the modern Himalayas, with beautiful rivers flowing in the valleys nearby. Then, once we were on the Walker County side of the mountains, the swamp forests would stretch as far as the eye could see. Being tropical, they would be rain forests, but

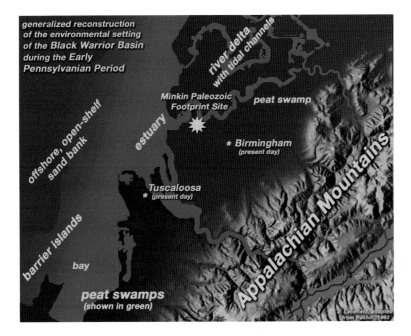

FIGURE 5.3 Paleogeographic setting of the Minkin Site area during the coal age. Note the current locations of Tuscaloosa and Birmingham, which are ~100 km (~60 mi) apart.

they would look nothing like the Amazon, the modern world's largest rain forest. The plants seen in the Amazon today would simply not have existed. Instead, the view on our flight over the coal-age swamps would be dominated by huge trees covered in scales.

As our plane approached the location of the Minkin Site, we would start to see the shore of the Black Warrior Sea, as well as a river draining into it, forming a large estuary. In the estuary, near the swamp forest, we would see a flat expanse of open mud (a *mudflat*) that would come and go with the cycling tides. That is, sometimes the mud would be exposed, and other times it would be covered with water. The effects of the tides would depend on the relative locations of the Earth, moon, and sun, since tides are caused by differential gravitational forces. There is evidence in the rocks at the Minkin Site that the mud never really had a chance to dry during low tide, implying that there could have been a persistent shallow film of water even at that time. The scene would not be exactly like Mobile Bay, a large shallow embayment on the southern Alabama coast, but more like the Mobile delta, where a number of rivers come together and drain into the bay.

Swamp Forest

There is only so much we could see from the air during our imaginary visit. If we want to experience life during the Pennsylvanian Period, we need to find a place to land and then go take a look around. Let us imagine landing near the ancestral Minkin Site, and then walking into the nearby forest. Figure 5.4 shows a beautiful painting from the Field Museum of Natural History in Chicago that accurately depicts what we might see. Virtually every kind of plant shown in this picture has been found in fossil form at the Minkin Site. In the foreground the large trees whose trunks are covered with diamond-shaped scars are the arborescent lycopods whose fossil bark impressions are called *Lepidodendron*. Farther back in the middle of the painting we see the huge trunk of another lycopod, whose fossil bark impressions are called *Sigillaria*. Extending outward from the base of the *Sigillaria* trunk are long roots that in fossil form are called *Stigmaria*. Although the tops of the foreground lycopods are mostly cut off in the painting, these tops would have included all the branches of these trees, and up close they would have had a finely detailed, hairlike appearance (chapter 17). In the foreground to the right in the painting, we see a different kind of tree related to modern horsetails. The most common fossil related to this type of tree is called *Calamites*. Finally, the forest floor is littered with smaller plants interspersed with seed ferns, one example of which is shown in the left middle of the painting. These tree-sized fernlike plants, represented by foliage like *Neuralethopteris* and large seeds such as *Trigonocarpus*, have no modern relatives; their cup-shaped pollen organs (*Whittleseya*) hung down like tiny bells from the branches. The plant assemblage would have included a variety of smaller true ferns (which reproduce by spores).

The scientist who has studied Minkin Site plant fossils in the most detail is Dr. David L. Dilcher, a former graduate research professor (now retired) at the Florida Museum of Natural History and a distinguished member of the US National Academy of Sciences. The specimens he inspected were collected by a variety of people from mine "spoil piles," but the context of those specimens (that is, which piles they came from) was not recorded. He concluded that the flora was reasonably diverse and similar to what he had seen in other mines he had visited in Ohio,

FIGURE 5.4 Artist's rendition of a coal-age forest.

Europe, and the Illinois coal basin. However, in those other places he had noticed that different plant fossils—like the pith casts of the horsetail (*Calamites*), the giant arborescent lycopods (for example, *Lepidodendron*), the ferns, and the seed ferns (Pteridosperms)—seemed to be found in different horizons (layers) or different areas. He also suggested that some of the seed ferns from the Minkin Site might prove, on further study, to be unique to the Black Warrior Basin (D. Dilcher, pers. comm. 2010).

According to Professor Dilcher, some of the fossil plants collected at the Minkin Site are typical of higher and drier places in coal swamps. He notes that the lycopods would have lived in the wetter areas, growing in shallow water perhaps no more than a third of a meter (a foot) deep, like bald cypresses. *Calamites*, on the other hand, likely grew in somewhat drier areas that would have been on higher ground, small natural hills called hammocks. The difference in elevation might have been less than a meter, but that can make a huge difference in wet areas.[8]

At any active surface coal mine in Alabama, you are bound to see slabs covered with what looks like snake skin. The closely packed scales may be coal black; less well-preserved examples are simply impressions in shale or sandstone. There were no Pennsylvanian snakes, so what is this stuff? Not fish skin, which would not be made of carbonaceous black material. How about bark, the most recognizable remnant of the giant arborescent lycopods? The lycopods were some of the most remarkable plants of the period. They grew more than 30 m (100 ft) high and had trunks more than a meter (3 ft) in diameter (Lacefield 2000). Leaves grew directly from the tree trunks. As the trees grew, the chute-like leaves were shed, and the trunks became covered with diamond-shaped leaf scars that looked like scales. In fact, *Lepidodendron* means "scale tree." Although the actual color of these trees is not known, it is thought that the trunks were green, not the brownish gray of modern trees. This is deduced from the properties of the leaf scars.[9] According to Lacefield (2000), most of the biomass of Alabama coal comes not from *Lepidodendron* but from the related form genus *Lepidophloios*, whose scars are more similar to fish scales than are those of *Lepidodendron*. These distinctions are highlighted further in chapter 17, which includes examples of these as well as other plant fossils from the Minkin Site.

What happened to these magnificent trees? They are thought to have gone extinct by the end of the Paleozoic Era, probably because the swampy conditions needed for their survival eventually faded away, or because of competition from other plants. Small aquatic or semiaquatic plants called quillworts are the only known surviving descendants of these ancient giants.[10]

The Field Museum painting (fig. 5.4) contains illustrations of Carboniferous insect life. Because of the high atmospheric oxygen content, some insects achieved colossal size. For example, a Carboniferous dragonfly ancestor could have a wingspan of nearly half a meter (1.5 ft). We know such insects lived at the Minkin Site because fossils of several large primitive wings have been found there (chapter 16). The painting shows a large dragonfly fluttering into the coal-age forest. It also shows a cockroach crawling on a lycopod trunk in the left foreground. The typical Carboniferous roach was no bigger than those of today (Snodgrass 1930). There is no evidence in the trace fossil record for roaches at the Minkin Site, which suggests that they stayed in the forests and did not go out onto the mudflats at low tide.

The only major groups of terrestrial arthropods to leave numerous trackways at the Minkin Site were primitive monuran (wingless) jumping insects that are relatives of modern bristletails, and primitive millipedes. Both of these could achieve sizes of five centimeters or more, and they are described further in chapter 16.

MUDFLAT

While terrestrial animals surely would have lived in the forest, and some were undoubtedly big enough to eat the large insects that lived there, smaller ones may have preferred to eat what was crawling out on the mudflat. Or, perhaps the animals took advantage of low tide to quickly reach other parts of the forest and new venues for food. In either case, the mudflats could have been very active during low tide. What might we have seen during our imaginary visit?

In order to answer this, we first need to examine how tracks were preserved at the Minkin Site. A mechanism of deposition is critical; that is, there has to be water and there has to be a significant source of small particles that can bury a specimen (be it a plant or a footprint) relatively quickly (Pashin 2005). Interpreting what we would have seen on the mudflat is complicated because different layers of rock represent different times. Five different tetrapod trackway species identified at the site (chapters 14 and 15) suggest that amphibians and early reptiles coexisted there. Yet in very few cases was a slab found that included, on the same horizon (rock layer surface), trackways of amphibians and reptiles (or even of different reptile ichnotaxa) that could have been made at about the same time.[11] Also, of the several invertebrate trace-fossil species, including those made by monuran insects, millipedes, and horseshoe crabs, the only one commonly found on the same slabs as vertebrate trackways is *Treptichnus apsorum*, the most abundant Minkin Site ichnospecies, which consists of burrows made by fly larvae or something similar.

The Pennsylvanian water at the Minkin Site was most likely freshwater flowing out of a river valley. Modern amphibians, millipedes, and insects live only in freshwater environments, and this was presumably true for the ancient tracemakers. Even horseshoe crabs, or xiphosurans, makers of the trace fossil *Kouphichnium* at the Minkin Site, do not necessarily indicate a marine (seawater or saltwater) environment. A. Rindsberg (pers. comm. 2010) noted that some Carboniferous xiphosurans (such as *Euproops*, in the Mazon Creek fauna) lived as adults in a freshwater Carboniferous equivalent of mangrove swamps. While modern xiphosurans (limulids) spend their adult lives almost entirely in fully marine water, they can enter brackish (slightly salty) estuaries to breed.

Mountain building and climate were the main factors controlling deposition in the Black Warrior Basin. As noted previously, cyclic climatic factors, glaciation in the Southern Hemisphere being one of the most important, caused sea levels to rise and fall over thousands and millions of years. When sea level falls, land is exposed and eroded, and deposition shifts to lower areas. When sea level rises, land is flooded, creating what stratigraphers call "accommodation space." The principle is simple. Most sediment is deposited by water, and water can deposit sediment only under its surface. So when the sea rises, lagoonal mud, barrier island sand, river deltas, and so forth get deposited along the coasts. Deposits made by the old cycle are buried, and the cycle begins again. This process of accumulation of sediment layers on the coastal plain is also called *aggradation*, which could be loosely translated as "upward accumulation" or "upward grading."

The sources of sediment in the Pottsville Formation were to the southwest, in the Ouachitas (Cleaves 1983; Thomas 1988), and eastward and northeastward, in the Appalachians (Pashin and others 1991; Pashin 1993). The very mountains whose weight made the basin sink were being eroded by rivers, and their tiny broken bits went to fill the basin up. Geologists talk about the rock cycle, in which preexisting igneous, metamorphic, and sedimentary rocks are worn away and made into new sedimentary rocks. That is basically what the Black Warrior Basin was all about and, because this sediment was deposited in swampy environments that were teeming with plants, is the reason there are coal mines in Alabama.

The active erosion of the mountains surrounding the Black Warrior Basin provided fine-grained sediment that was very effective at recording the activities of animals walking on the mudflats. We can imagine the following: at low tide, the water receded, leaving mostly wet mud over a large area (possibly many square miles) and perhaps shallow puddles or pools of water here and there where an occasional fish would get trapped. The trapped fish were forced to skim the bottom with one or more of their fins (producing the trace fossil *Undichna*; chapter 15). During this time, whatever terrestrial creatures lived in the area would go out onto the mudflat either to look for food or to move around for other reasons, leaving tracks or traces in their wake. Insect larvae would have been abundant on the flat and could have provided some of this food, because many fossil traces of these larvae have been found at the Minkin Site (*Treptichnus, Arenicolites*). As the animals moved, they would have produced both surface tracks and undertracks (or traces). Small amphibians and reptiles could go out only when the water was shallow or mostly gone, while larger ones could walk and leave footprints in somewhat deeper pools. Presumably, in our imaginary time travel we would not see amphibians and reptiles happily mingling out on the mudflat because we simply do not find much evidence for it in the trace fossils collected at the site. At one time we might see amphibians, at another time reptiles, at another time larger animals, and at still another time large numbers of jumping insects or horseshoe crabs. But we would not necessarily see all of these on the same day or in the same place.[12]

If at low tide, significant swaths of mud have a chance to dry, there is a possibility of mud cracks developing. These can be preserved as fossils just as footprints and other traces are. However, no mud crack fossils have been found at the Minkin Site. Surface tracks at low tide were made in mud that never dried. Naturally, when the tide did come in, the water would have washed away surface tracks, especially where the mud was very soft. But as noted in chapter 3, the preservation potential of undertracks is higher because they are already buried and partly protected. The way to preserve these tracks, then, is to continually deposit new sediment on the mudflat and bury them deep enough that when sea level falls, the track-bearing layers are not destroyed by erosion before the next cycle of forest-building can start. The sediment content of the water during the Westphalian A must have been exceptional. The Mary Lee coal zone is the only one in the entire Pottsville Formation that preserves fossil animal trackways (J. Pashin, pers. comm.).

Figure 5.5 shows an artist's rendition of the Minkin Site during the coal age. We see a dense forest of coal-age plants near a shoreline, with a large anthracosaur wading in the shallow water. These creatures are thought to have left the large tracks called *Attenosaurus subulensis*. They were the only trackmakers big enough to do so.

FIGURE 5.5 Artist's depiction of the Minkin Site during the coal age. The large, towering trees with all their leaves at the top are the giant scale trees known as *Lepidodendron*.

In summary, our visit to the Pennsylvanian Period in Alabama reveals a world of very high mountains, vast expanses of tropical swamps filled with strange trees and oversized insects, a huge inland sea, a warm and humid climate without real seasons, likely active volcanoes and frequent earthquakes, and large numbers of tetrapods. The stage was set for the long reign of the reptiles and the eventual coming of the Mesozoic Era: the Age of Dinosaurs. However, the land, the animals, and the climate were not the only things that were different at the time. We have been looking only down: at the ground, the scenery, the forest, and the water. We have not looked up. Let us now do so.

6

The Pennsylvanian Period in Alabama: Looking Up

The previous chapter took you on an imaginary journey back to coal-age Alabama, specifically to the Minkin Paleozoic Footprint Site at the time when its trackmakers were still alive and active. Looking around, we would have seen a beautiful mountain range, a dense tropical forest, a large river, and the characteristic plants and animals of the time. All these things would have been very different from what we see in Alabama today. But an unfamiliar landscape is not the only thing we would notice. If we waited for nightfall, we would also observe a sky full of unfamiliar stars. There would be no constellations we would recognize, and even the Milky Way itself would look different. If the Pennsylvanian daytime world seems alien to us compared to our modern world, the night sky of the Pennsylvanian world would really make it seem like we were on another planet.

The reason the night sky of Pennsylvanian Alabama would be unrecognizable has to do with both the way stars move in our galaxy and how stars change over time. These changes are by and large inconsequential over a human lifetime (or even a hundred human lifetimes) but over a period of 313 million years would be very significant. In order to provide a more complete picture of the world of the Minkin Site trackmakers, this chapter focuses on the night sky and what it, the Milky Way, the planets, and other celestial objects might have been like when these animals were alive.

One of the most remarkable things about the night sky is that it is like a time machine. When we look out into a starry night, we are looking back in time.[1] Any events we observe in the sky, such as the explosion of a star, are always like old news: the event happened long ago, but we are receiving news of it only now. This happens because light, which carries the news, travels at finite speed, 299792.5 km/s (~186,000 mi/s). In a year, a beam of light travels one light-year, or 9.6 trillion km (6 trillion mi). This means that the farther out in space we look, the farther back in time we see. For example, Sirius, the brightest star in the night sky, is 8.6 light-years away and appears to us as it did 8.6 years ago. The Orion Nebula, on the other hand, is 1,344 light-years away[2] and appears the way it looked in about 670 AD. The Andromeda Galaxy, the most distant object visible to the naked eye, is 2.5 million light-years away, appearing to us the way it looked long before our species evolved.

These light-ages pale in comparison to the time of the Minkin Site trackmakers. Sunlight that shone on the mudflat and reflected back into space is now more than 310 million light-years away. Out at that distance, we find the famous Coma Berenices galaxy cluster, Abell 1656 (fig. 6.1), which is 310 million (plus or minus 22 million) light-years away.[3] Coma is one of the most important concentrations of matter in the nearby universe, including more than a hundred trillion solar masses of bright and mostly dark matter, and thousands of individual galaxies (Colless 2001). The cluster has been the subject of much significant astronomical research, but for us it is sufficient to know that within the uncertainty of the distance, the light we are receiving from these galaxies today left them during the Westphalian A. Whatever we see happening in these galaxies, or whatever measurements we make on them, is all very old news.

The star clouds forming the Milky Way tend to be a few thousand light-years away. At certain times of the year, the Milky Way crosses the sky like a sparkling river, its beautiful star clouds interrupted by giant clouds of interstellar dust. It may be surprising, but "dust" is one of the most common substances studied by astronomers. Dust is thought to be the "smog" of stellar evolution, material made of heavy chemical elements that were processed in the centers of stars or produced in the violence of stellar explosions. Dust is thought to condense in the atmospheres of evolved stars that eventually eject their outer gases back into

interstellar space. After a long time, the dust of many generations of stars has collected in the plane of the Milky Way, where it blocks our view of what lies behind it.

The Milky Way is the backdrop for the history of our solar system. It is our home galaxy, a highly flattened system of stars, gas, dust, and other forms of matter bound together by gravity into a single physical unit. It is believed to be approximately 100,000 light-years in diameter, larger than the average galaxy but typical of many bright galaxies. The brightest part of the Milky Way is in the direction of the constellation Sagittarius. This way lies the center of our galaxy, a point believed to harbor a supermassive black hole four million times as massive as the sun (Ghez et al. 2008), and to include many thousands of stars in a region only a few light-years across. In the absence of interstellar dust, the central nucleus of our galaxy would be brightly visible to the naked eye on warm summer nights in Alabama. However, there is so much dust in between that the galactic center is completely hidden. To see it, we must observe it with telescopes sensitive to longer wavelengths of light than visible light.

FIGURE 6.2 The inclined spiral galaxy NGC 3953 in Ursa Major, a galaxy that likely resembles ours. This image is similar to the way our galaxy would look if it could be viewed from outside its plane. The *X* is intended to give a general idea of where the sun would be relative to the center if NGC 3953 were our galaxy.

FIGURE 6.3 The edge-on spiral galaxy NGC 4565, which resembles our perspective on our own galaxy. This one is viewed from far outside, but our perspective is from inside the Milky Way.

The galactic center is important because all objects in the disk of the Milky Way rotate around that point. If the Milky Way could be seen from the outside, it might resemble the galaxy shown in figure 6.2, known as NGC 3953, in Ursa Major. NGC 3953 is a spiral galaxy, a massive pinwheel-shaped system whose stars and interstellar material have been molded into spiral arms through the action of gravity. Although the arms of NGC 3953 are visible to us because of its more face-on orientation, we cannot see the arms of the Milky Way directly. We infer them from detailed observations. Our perspective on the Milky Way is an edge-on, inside view, and as shown by the galaxy NGC 4565 in figure 6.3, spiral arms are not detectable in such a view. Instead, we see the central bulge poking out on either side of a plane of dust running along the axis of the galaxy.

It takes a long time for the sun to orbit the galactic center once. Recent studies have indicated that the sun and its retinue of planets, moons, asteroids, and comets orbit at a distance of 27,400 light-years from the galactic center at a speed of 820,000 km/hr (512,000 mi/hr) (Ghez et al. 2008). The sun orbits far enough from the center to keep us safe from the massive central black hole, but it is so far out that we cannot characterize the sun's orbital period in mere decades or centuries. At this distance, it takes nearly 240 million years for the sun to go around once! This number corresponds closely with the beginning of the Mesozoic Era, the Age of Dinosaurs. Thus, one "galactic year" ago, when the sun was in roughly the same area (relative to the galactic center) as it is now, the Age of Dinosaurs began. The dinosaurs, one of the most successful groups of land animals that ever lived, lasted about three-quarters of an orbit. The Age of Humanity began so recently that the sun has moved only a few degrees along its orbit since our ancestors walked in ancient Africa.

We can visualize some of this with the map in figure 6.4, which is a schematic representation of the sun's orbit in the plane of the Milky Way. The Minkin Site fossils are about 1.3 galactic years old. With modern data, we can map out the approximate appearance of the sun's orbit and indicate where our solar system was located 313 million years ago. Figure 6.4 shows that the sun's orbit is roughly circular but does not close on itself exactly.[4] The orbits of planets around the sun are mostly closed ellipses, which happens because the sun is much more massive than any other body in the solar system, and because the planets orbit far from the sun. But the sun orbits inside the Milky Way, meaning there is considerable matter outside the sun's orbit. In such a circumstance,

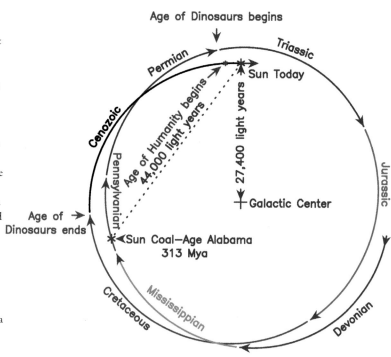

FIGURE 6.4 Geologic time in terms of the sun's galactic orbit. The current location of the sun is at the asterisk labeled "Sun Today," while the asterisk labeled "Sun Coal-Age Alabama" is approximately where the sun was located when the Minkin Site tracemakers were alive. The arrows mark the boundaries of the various periods of Earth's history. The Carboniferous Period is shown separated into the Pennsylvanian and Mississippian Periods. These are part of the Paleozoic Era, as are the Devonian and Permian Periods. The Mesozoic Era is shown as the Triassic, Jurassic, and Cretaceous Periods. The Cenozoic Era is shown undivided but is mostly the Tertiary Period, with the Quaternary Period beginning around the time of the "Age of Humanity."

we do not get the usual perfect ellipse, but an orbit that never repeats itself. The schematic shows that one galactic year ago, the sun was actually farther from the galactic center than it is now, by about 2,800 light-years. The schematic also shows that 313 million years ago, the sun was located about 44,000 light-years ("Sun Coal-Age Alabama") from where it is today ("Sun Today"). Note that figure 6.4 does not account for the fact that the sun also oscillates slightly in and out of the plane of the Milky Way as it orbits or for the effects of structure in the galactic disk, such as spiral arms.

As the sun moves along its orbit, the Earth's perspective on the universe changes. What could we have seen in the sky over the Minkin Site at night, 313 million years ago, when the area was teeming with primitive tetrapods and arthropods? For example, would we have been able to see the constellation Orion, with its familiar bright stars like Betelgeuse and Rigel, or the popular asterism known as the Big Dipper?[5]

During much of recorded human history, the stars have seemed fixed in position and mostly unchanging in brightness or color. We see basically the same constellations that the ancient Egyptians saw 5,000 years ago. But immutability is an illusion. If we were to follow the course of the sky over a period of 313 million years, we would see drastic changes. Imagine the sky from this time up to the present as a video where each frame is a snapshot taken every 100,000 years, and which, when played, advances one frame every second. The video would then run for 3,130 seconds, or 52 minutes. While watching the video, we would see stars moving in all directions; the sky would be filled with movement. Some stars would get brighter as they approached us, and others would get dimmer as they receded. Some stars would, relatively speaking, "pop" into view and before you knew it would intensely brighten and then disappear. Other things would come and go, too, like the bright nebulae surrounding new stars and the nebulae surrounding dying stars. The "walk of stellar life" would be in full view, and it would seem random.

These changes would occur for two different reasons. The first is that the Milky Way does not rotate like a solid object. Relative patterns on a solid object are maintained as the object rotates (as, for example, the longitudes and latitudes of cities as Earth rotates), but in the Milky Way, stars that are farther from the center take a longer time to go around than stars closer in. Galactic rotation is therefore more like that of a fluid whirlpool than that of a solid object. Also, orbits are generally not closed relative to the galactic center (for example, the sun's orbit in fig. 6.4). This changes all relative positions. Therefore patterns, such as constellations, also change their aspects over long periods.

Figure 6.5 shows the changes in the Big Dipper that have occurred and will occur. While the ancient Egyptians saw basically the same Big Dipper that we see, one hundred thousand years ago the pattern would not have been called a "dipper." One hundred thousand years from now the pattern will change yet again into something that also would not be called a dipper. The Big Dipper is unusual (compared to other star groups visible to the naked eye) in that five of its seven stars are moving in space together, so all the pattern change is due to two unrelated stars. If such changes as can occur for the Big Dipper can become noticeable over a period of only a hundred thousand years, imagine the changes that would occur over 313 million years!

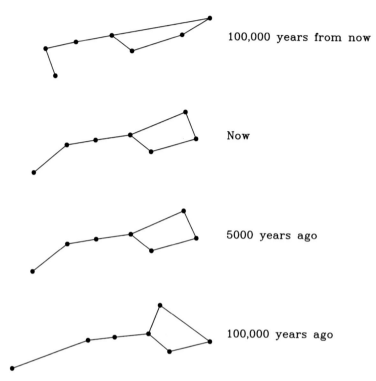

100,000 years from now

Now

5000 years ago

100,000 years ago

FIGURE 6.5 The appearance of the Big Dipper in the past, present, and future, showing changes that can occur even over the relatively short time span of 100,000 years. The appearance of constellations cannot be extrapolated too far forward or backward because the long-term movements of stars are not known with sufficient precision.

As an example, consider Rigel Kent,[6] currently the nearest star (it is a system of three), 4.3 light-years away, and the third brightest star in the night sky. It is thought to be almost five billion years old, roughly the same age as the sun.[7] Where was it 313 million years ago? Although Rigel Kent's orbit is close to ours, its distance from us and its direction in the sky vary considerably over time. Using its known orbital parameters, we have deduced that Rigel Kent was 2,900 light-years away and not visible to the naked eye during the Westphalian A. Extrapolating back we discover that Rigel Kent is closer to us now than it has ever been, at least since the Devonian.

The second reason the night sky would change noticeably over a period of 313 million years is the relatively short lifetimes of massive stars. A star like the sun can shine steadily, and provide a stable source of heat and light, for almost ten billion years, deriving its luminosity from the fusion of hydrogen into helium in the center. Stars that are powered this way are called main sequence stars. Our sun is a main sequence star with a radius of about 700,000 kilometers and a surface temperature of 5800 degrees Kelvin. It is probably 1 to 2 percent brighter now than it was 313 million years ago. Main sequence stars that are much cooler than the sun are only a few tenths the sun's radius and are very dim compared to the sun, while the most massive main sequence stars can be ten to fifty times the sun's radius and a million times as bright.

The main sequence lifetime of any star is strongly dependent on its mass. Hot stars more massive than fifteen solar masses (meaning the mass is fifteen times that of the sun) have main sequence lifetimes of less than 10 million years. In our imaginary sky video, the entire lifespan of such a star would be over in less than two minutes. In contrast, viewed in the same way, the sun's lifespan would be a little more than a day, much longer than our 52-minute video. Stars more massive than about eight to ten solar masses are believed to end their lives violently in a titanic explosion called a *supernova*. Less massive stars die more peacefully by ejecting their outer gases more slowly. In either case, only a dim or virtually invisible remnant of a star's former glory is left behind, and it then fades from view.

A four-solar-mass star has a main sequence lifetime equal to the age of the Minkin Site, 313 million years.[8] This means that any currently visible star more massive than four solar masses did not exist during the Westphalian A. This is true of all the bright stars in the constellation Orion: they are eight to forty times the mass of the sun.[9] The red supergiant Betelgeuse is probably the oldest of them all, at 10 million years.[10] Thus, we could not have seen Orion 313 million years ago because none of its main stars even existed at that time.

The stars making up the constellation Orion are exceptionally bright but are also extremely rare. They are prominent in our sky only because they are so luminous that they can be seen from great distances. In fact, the stars that make up all of our constellations, not just Orion, are generally like this. For example, the bright star Antares in the constellation Scorpius is more than 600 light-years away. At such a distance, the average star in the Milky Way would be well below naked-eye visibility. Most of Orion's bright stars will probably end their lives as a supernovae. In our imaginary video, we would have to catch such an explosion at precisely the right moment to notice it.

Most stars in our galaxy have main sequence lifetimes longer than that of the sun, and many are as old as or older than the sun. Again, the Milky Way's whirlpool-like rotation means that the relative positions of older stars would change.

Star clusters are important groupings of stars that have formed as a family. The most prominent of such clusters is the Pleiades (M45 in Taurus, the "Seven Sisters" of mythology), located a mere 400 or so light-years away and easily visible to the naked eye on fall and winter nights in Alabama. From studies of star brightness and color, we can estimate the age of a star cluster like the Pleiades, assuming all of its stars formed at about the same time. In this case, we discover that it is only 100 million years old. The stars in the Pleiades were not shining in anyone's sky 313 million years ago. Half the clusters like the Pleiades in the Milky Way disintegrate (that is, lose all their stars) in less than 200 million years (Wielen 1985).

Our sun could have been born in a cluster. When very young (less than 10 million years old), these clusters show pink glowing gas and dust, the color being due to hydrogen gas excited by ultraviolet light from massive stars. Pink nebulous clusters line the spiral arms of the distant galaxy, the famous Whirlpool Galaxy M51, shown in figure 6.6, but none of those seen in the figure existed 313 million years ago. The figure shows the way the galaxy looked 26 million years ago. If we could see M51 as it is now, most of the pink nebulae would be gone, and new ones would have sprung up in other locations. In general, pink nebulae tend to concentrate along spiral arms. If the Milky Way is mostly a two-armed spiral, then the sun would pass through an arm every 100 million years or so. At these times, the sun might pass near one of these nebulae, which look the way they do only because of the presence of very massive stars. This presents the possibility of the sun being close to a supernova every 100 million years or so.

Particularly interesting is how far a whole galaxy can move in 313 million years. For example, the Great Spiral in Andromeda, currently 2.5 million light-years from Earth, is approaching our galaxy at 300 km/sec. This may not sound very fast, but since the Minkin Site trackmakers

Figure 6.6. Color image of the Whirlpool Galaxy M51 in Canes Venatici. The spiral arms are lined by pink nebulae, which are clouds of hydrogen atoms excited into a glowing state by hot, massive (and short-lived) stars.

were alive, the Andromeda Galaxy has moved 313,000 light-years closer to us. This is three times the size of the Milky Way and means that Andromeda would have appeared a little fainter to the naked eye when the Minkin Site trackmakers were alive.

A famous star in our night sky is Polaris, otherwise known as the North Star because it lies in the direction of the north celestial pole. Owing to precession of the Earth's rotational axis, the pole completes a circle on the sky once every 26,000 years. Thus, 26,000 years from now, Polaris should once again be the North Star. But Polaris could not have been the North Star 313 million years ago, because it has a mass of 7.5 solar masses.[11] It didn't exist!

Although the planets certainly existed 313 million years ago, they might have looked different. The moon would have looked about the same as it does now, but the lunar crater Tycho was not there. It has an estimated age of only 108 million years.[12] Tycho is easily the moon's most visible impact crater because of its high brightness at full moon and its extensive ray system. Its absence would slightly change the naked-eye appearance of the moon. The moon was likely already tidally locked into keeping the same face toward the Earth by Pennsylvanian time.

The latest studies suggest that at least some of Saturn's rings are very old.[13] But they are constantly changing and are kept fresh by interactions with the planet's many moons. They could have looked different 313 million years ago. The rings were once thought to be only 100 million years old. Jupiter's Great Red Spot has persisted for 350 years. Because Jupiter has a completely fluid surface and a lot of internal heat, there is nothing to weaken the storm. Would the spot have existed 313 million years ago? Smaller storms come and go regularly.

Mars has a thin atmosphere and regular weathering, but it would have been just as cold and dry 313 million years ago as it is today. Any open bodies of water it might once have had would have been long gone by then. Venus probably evolved rapidly after forming, but 313 million years ago it was already in a very slow phase of evolution. It would have looked much the same as it does today.

As for the Earth, we have already noted the collision of continents that uplifted the Appalachian Mountains in Pennsylvanian time. The only major continent that existed at the time was the supercontinent Pangaea, so the Earth would have looked much more different compared to today than the moon would have looked. Because the moon and the Earth are relatively close to each other, back then, as today, ocean tides would have occurred. However, these ocean tides do more than simply affect sea levels. When combined with the rotation of the Earth, they also cause friction that slows the rotation of the Earth by 1.7 milliseconds per century.[14] This has been inconsequential during human history, but if we extrapolate this very small effect back 313 million years, we discover that the "day" during Pennsylvanian time was only 22.5 current hours long. The same tidal forces have been pushing the moon farther away from Earth at a rate of 2.2 cm/yr.[15] The moon was 6800 km closer to us 313 million years ago, but still not close enough for us to see craters easily with the naked eye.

In summary, the world of the Minkin Site trackmakers was different from that of today in more ways than climate, latitude, landscape, and atmospheric oxygen content. The sun was slightly dimmer, the day was more than an hour shorter, the moon was a little closer while the Andromeda Galaxy was a little farther away, none of the present constellations existed, the sky had bright stars that have long since gone to the stellar graveyard, familiar star groups like the Pleiades or the bright stars in Orion had not yet been born, any pink nebulae the sun was near have largely disappeared, Rigel Kent was not the nearest star to the sun and was not even visible to the naked eye, Polaris could not have been the pole star, and the moon was missing its most prominent crater. The Earth's whole perspective on the rest of the universe was different because the sun was in a completely different location relative to the galactic center. The world of the Minkin Site trackmakers was indeed alien and unfamiliar.

7

Carbon Hill Redux

Chapter 1 described some of the story behind the first discovery of fossil tracks in Alabama. These were found in the 1920s in an underground coal mine near Carbon Hill, in western Walker County. Carbon Hill started as a small mining town in the 1880s, just a few years before R. E. Galloway bought property in the area and successfully expanded his mining business.[1] As the name implies, coal is important in the area. Reclaimed surface mines dot the landscape, and active mining continues to be part of the economy, as it has been for more than a century.

As we described in chapter 1, the fossil track discoveries near Carbon Hill were documented in a paper titled *Footprints from the Coal Measures of Alabama*, by Truman Heminway Aldrich Sr. (1848–1932) and Walter Bryan Jones (1895–1977), published as Alabama Museum of Natural History Paper 9 (1930). Reprints of the paper languished unsold for decades until the Union Chapel discoveries. Here we elaborate a little more on the discovery and further our acquaintance with the people involved.

Aldrich (fig. 7.1, top left) was a mining engineer whose hobby was shell collecting.[2] In a remarkable career spanning banking, coal mining, and even a brief stint in Congress, Aldrich began to study Tertiary Period fossils (2.4 to 65.5 million years ago; fig. 2.1*B*) after he met Dr. Eugene Allen Smith, the distinguished state geologist of Alabama (Gardner 1933, 301). Aldrich was so talented at collecting and recognizing fossils that Smith eventually appointed him curator of paleontology at the Alabama Museum of Natural History. On a sheet of museum stationery dated July 31, 1929, Aldrich is listed as "Hon. T. H. Aldrich,

FIGURE 7.1 *Top left*, Truman Heminway Aldrich (1848–1932); *top right*, Walter Bryan Jones (1895–1977). *Bottom left*, W. Frank Cobb Sr. (1894–1966), general superintendent of the Galloway Coal Company in the 1920s when the first tracks were discovered. T. H. Aldrich named the most common tracks found at the No. 11 Mine *Cincosaurus cobbi* out of gratitude to Mr. Cobb. *Bottom right*, Arthur Blair, geologist of the Tennessee Coal, Iron, and Railroad Company of Birmingham at the time of the first trackway discoveries at the No. 11 Mine; he was the first professional geologist to visit the No. 11 Mine and verify the discoveries.

Paleontology." By October 15, 1931, museum stationery listed him as "Dr. T. H. Aldrich,"[3] reflecting the honorary Doctor of Science degree he had received from the University of Alabama for his "great service in the industrial and intellectual development of the State" (Gardner 1933, 301).

Jones (fig. 7.1, top right) was the Alabama state geologist at the time of the trackway discoveries and served in the role from 1927 to 1961. The main Geological Survey of Alabama building on the University of Alabama campus is named after him. He was the protégé of Eugene Allen Smith and earned considerable distinction in his career, which included the preservation and study of Moundville Archaeological Park. At the time of the trackway discoveries, Jones was in his thirties while Aldrich was an octogenarian. In an undated letter attached to one dated October 15, 1931, Jones states, "The Aldrichs' are going up with me in the morning. The office got Mr. Aldrich a case of Edgeworth [pipe tobacco] for his birthday and all marched into his office for the presentation. He was much pleased. His birthday is tomorrow [October 17]."[3] This was Aldrich's last birthday celebration. He died on April 28, 1932.

The No. 11 Mine of the Galloway Coal Company, which was in operation near Carbon Hill until about 1934,[4] was a huge underground mine covering an area of about 8 km^2 (3 mi^2), laid out in the "rooms and pillars" style, in which underground spaces were cleared or mined of coal, but pillars of unmined coal were left in place to support what was above. The mine consisted of hundreds of these coal-excavated chambers, according to a detailed map made in 1935.[5] The schematic in figure 7.2 shows an outline of the mine and a few reference points from this map that are helpful in interpreting locale information given by Aldrich and Jones. The entrance was centered within a cluster of buildings on the north side of the mine, including a supply house, a machine shop, a boiler, a bath house, and a hoist house. Between the bath house and the supply house was a manway for walking entry into the mine. Once inside, one entered a series of "neighborhoods" with side streets coming off larger "thoroughfares." The mine is no longer accessible from the surface.

The main entrance near the blacksmith and car building led down a tunnel called "Main South." Branching off Main South were major tunnels labeled as right or left with a number, such as 3rd Right, 1st Left, and so forth, the "right" or "left" reckoned for movement to the south. The 7th Right entry was especially long and was labeled the "7th Right Slope." About a third of the way along this tunnel, a major branch called the "Southwest Slope" began. Most of the fossil trackways Aldrich and Jones described were found in this part of the mine, some along the "4th

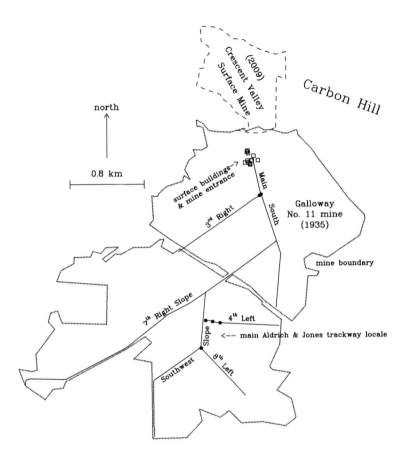

FIGURE 7.2 A schematic map of the No. 11 Mine of the Galloway Coal Company, an underground mine. The map is based on a 1935 original map likely made shortly after mining operations ended. The first trackways in Alabama were found along the "Southwest Slope," approximately 2.1 km (1.3 mi) from the mine entrance. The mine boundary is shown by the dotted outline. Several special underground tunnels (such as the "4th Left Entry," cited as a primary locale for many of the main tracks illustrated by Aldrich and Jones) are indicated. The filled circles along the "3rd Right," "4th Left," and "8th Left" branches show where specific tracks from Aldrich and Jones (1930) were found.

Left" branch. Others were found near the "3rd Right Entry" off Main South. To get to the 4th Left entry point would have involved about 2 km (1.3 mi) of walking underground. Aldrich and Jones noted that *Cincosaurus cobbi* trackways were "found throughout the mine" and that the animal that made the tracks was "undoubtedly the most abundant of all the reptiles living at that time." The specimens they illustrated in Museum Paper 9 were "from several places in and near the southwest slope." The most important of these specimens is shown in figure 7.3.

The first account of the No. 11 Mine tracks actually preceded Museum Paper 9 by a few months. At the request of W. Frank Cobb Sr. (fig. 7.1, bottom left), the Galloway Coal Company general superintendent, Jones wrote a one-page report on the discovery in the February

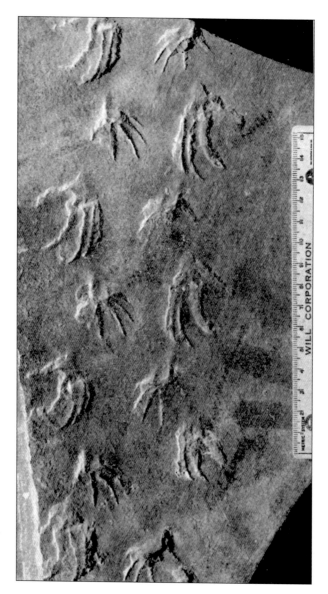

FIGURE 7.3 The main set of tracks that Aldrich named *Cincosaurus cobbi*, positive hyporelief from the 1930 report by T. H. Aldrich and W. B. Jones. This specimen shows how *C. cobbi* got its name: both the manus (smaller, inward-turned tracks) and the pes show five clear digits, indicating that the tracemaker was not an amphibian.

1930 issue of *Coal Age*, a popular-level magazine that covered everything about the coal mining industry, ranging from the newest mining techniques and equipment to the family, employee, and management issues of miners.[6] The magazine ran in printed form since about 1910 and in 1928 had published short articles about Galloway Coal Company mining in Alabama, including one on the No. 11 Mine and how the operation handled large displacement faults in the Jagger coal seam.[7] In Jones's *Coal Age* paper, he makes an interesting comment: while he credits Arthur J. Blair (fig. 7.1, bottom right) and Ivan W. Miller for the discovery of the tracks, since they were the ones who "brought them to light," he also states that he had personally known "for a decade" that such tracks existed. He does not repeat this claim in Museum Paper 9, only noting that workers had reported tracks "several years" prior to the discovery. Indeed, it is likely that the tracks were known even earlier, because mining began at the site at least as far back as 1912.[8] It appears to have taken the exceptional display of ceiling tracks seen by Blair and Miller to kindle strong enough interest in the tracks to bring them to the attention of the world. Jones states that Blair and Miller were "startled to behold the trail of the animal leading for some forty feet along the roof of the slope, until it disappeared into solid rock." In his *Coal Age* paper, Jones notes that the tracks defining the holotype that Aldrich named *Cincosaurus cobbi* (shown in fig. 7.3, named after W. Frank Cobb Sr.) were in fact the ones that Blair extracted from the ceiling display and took to his office at the Tennessee Coal, Iron, and Railroad Company (TCI) in Birmingham. When Jones saw these, he declared them to be "entirely new to Alabama geology." He later concluded that "the diversity of fauna was most surprising."

Jones comments in Museum Paper 9 about the cooperation of the coal company, and how the mine superintendent, H. M. Johnstone, supervised the removal of all specimens. These were not simply found in spoil piles. Jones wrote, "This work has been expensive and was carried out in a very careful, deliberate manner. The Company has fostered a notable contribution to science. Without this cooperation, the discovery could never have received the attention warranted by its importance." Almost to the letter, the same can be said about the Union Chapel Mine discoveries.

The main target of the mining operation in the Galloway Mine was coal from the Jagger seam, which in western Walker County is about forty feet below the Mary Lee coal bed, the main target of the Union Chapel mining operation. (In chapter 9, we argue that tracks in western

Walker County are from the same stratigraphic interval as those from the UCM in eastern Walker County, in spite of the coal seams having different names.) Tracks had been noticed in shale 0.8 to 1.1 m (2.5 to 3.5 ft) above the top of the Jagger coal seam. Jones noted that the tracks "are limited to slopes and entries which have been driven into the roof for additional clearance, or in caved roof." Jones observed tracks from several parts of the mine and pointed to "the fourth left entry off the southwest slope" as the best place for collecting. This area is indicated in figure 7.2.

Aldrich provided the "meat" of Museum Paper 9 by writing the detailed descriptions of the specimens and naming different ichnogenera and species based on the specimens that had been collected. As is fairly typical, the first specimens to be described were split into many different categories. For example, in addition to *Cincosaurus cobbi*, Aldrich recognized *C. fisheri* (named after the president of Galloway Coal Company, F. N. Fisher), *C. jonesii* (a name given to a specimen collected by Walter Jones), and *C. jaggerensis* (named after the coal seam just below the track-bearing layers). Of these four ichnospecies, only *C. cobbi* shows well-defined manus and pes traces with a full complement of five digits each. In hindsight one can see that the other three *Cincosaurus* ichnotypes as well as several of the other small track types named are likely preservational or behavioral variants of only one type: *C. cobbi* (Haubold et al. 2005a).

Excessive naming usually happens when too few specimens are collected to establish any basic relationships among apparent types. Aldrich and Jones were well aware of this, and in fact they were concerned about how effectively they could both present and interpret the material given their limited experience with tracks and the limited time they had to study them. In an unpublished paragraph connected with the paper, they make this rather remarkable statement:

> The names given to the specimens herein are to a certain percent fanciful, but they serve to identify the various species obtained from our Carboniferous rocks. They belong to the lower part of the Carboniferous series and are only a short distance above a good sized coal vein already mentioned. It is exceedingly difficult for a party who is not a trained anatomist to classify these footprints without

a great deal of time which we have been unable to give. We hope to be able to secure other impressions and also to connect these faint traces of life with other localities in Massachusetts, Ohio, Kansas, and Oklahoma with further study. We recognize this article as only a faint indication of what our Carboniferous series show. The writer is sorry that these descriptions are so indefinite.[9]

As we described in chapter 1, Walter Jones wanted to keep the discovery in Alabama, and this meant publishing something about it as quickly as possible.

It is customary in a scientific paper like Museum Paper 9 to present accurate information on where the main track types are discovered. For each specimen described (except *C. cobbi*, which was found throughout the mine), Aldrich gave the locality in terms of the mine "neighborhood streets," as we have described above. These can be interpreted only with a detailed mine map. The locales of the largest tracks found, *Attenosaurus subulensis*, are especially pinpointed, with descriptions like "southwest slope, 4th left entry, 188' from switch point," where the "switch point" may refer to the point where the 4th left entry breaks from the southwest slope (see filled circles in fig. 7.2).

One peculiarity of Museum Paper 9 is that it does not point in any way to the special visit of a famous paleontologist, George Gaylord Simpson of the American Museum of Natural History, to Alabama in late January 1930 to see the tracks firsthand and collect specimens to take back. The visit took place less than two months after Blair and Miller's discovery and more than two months before the manuscript of Museum Paper 9 was submitted. The only mention of Simpson is in an acknowledgment of him as the source of a complete bibliography of Carboniferous tetrapod footprints, included on the last two pages of the published document.

Simpson (1987) alludes to his Alabama visit in notes published as part of his collected correspondence, *Simple Curiosity*. In early 1930, he was on a largely southern tour of US sites, starting out from Washington to Florida, then to New Orleans, and back east again through Mississippi. On his way back to Washington, he stopped in Tuscaloosa, Carbon Hill, and Birmingham. His notes are ambiguous about what he did in Tuscaloosa and Birmingham, but on Carbon Hill they are very clear. He says: "A day spent weirdly and unsupported by the solace of tobacco in the depths of a coal mine, looking for the footprints of animals dead since 250,000,000 B.C." The date is based on an old calibration of the rock record. While he was in Alabama, and probably after his visit to Carbon Hill, he points to meeting a "very suspicious Professor of Geology in a tiny college who thought me a city slicker come to steal his lousy specimens." Given that this comment comes after that about the tracks, it is possible that Simpson was referring to a professor in Birmingham.

Simpson's return to New York was followed by a press release highlighting some of his finds. An article in the *Boston Herald* on January 29, 1930, states that the "most important result of Dr. Simpson's recent southern trip, according to the museum, was in obtaining numerous slabs with fossil footprints from a coal mine in Alabama and examining the conditions under which they occur. These tracks were recently discovered in a mine of the Galloway Coal Company at Carbon Hill, and were brought to scientific attention through the interest of A. J. Blair, geologist of the company, and W. F. Cobb, general manager. A series of representative tracks was obtained and will be exhibited by the museum after they have been studied and classified."[10]

In the *New York Times* article for the same day, the press release states that the "tracks occur in hard shale or slate two or three feet above the coal seam. When these deposits were being formed, in the Pennsylvania period, about 250,000,000 years ago, this country was flat and low and covered periodically with black mud. Across this mud walked numerous individual animals of that time. New layers of mud buried their footprints and preserved them so that with the passage of these millions of years, they are still preserved as clear impressions in the solid rock. They are among the oldest known traces of land animals."[11] The article goes on to describe the animals that made the tracks, recognizing them as tetrapods but suspecting that some were bipedal because several of the fossil trackways showed only hind feet impressions. It is possible that at this time, the phenomenon of undertracks was not fully appreciated. The article also alludes to the tracks being made when reptiles were newly evolving from their amphibian ancestors.

In addition to these reputable articles, tabloid newspapers of the time also reported on the tracks from Carbon Hill. In an amusing article dated May 18, 1930, in the *Portsmouth (OH) Daily Times*, two pictures from Museum Paper 9, one of *Attenosaurus subulensis* and one of *Cincosaurus cobbi*, are combined to make it look like two animals were actually engaged in some sort of violent encounter! The creatures were imagined to be huge—twenty-five feet long—and fighting to the death. The article claims that a skull and some bones of one of the fighters were also found in the mine.[12] The exact same article also appeared in the *Ogden (UT) Standard-Examiner*.[13]

The *New York Times* and *Boston Herald* articles indicate that George Simpson actually collected slabs of Alabama trackways for the American Museum of Natural History (AMNH). Naturally, we were very interested in finding out what he collected, what he might have written about those specimens, and whether the proposed exhibit at the AMNH was ever constructed and displayed. It turns out that Simpson did not write any research papers on the Alabama tracks, nor was there an AMNH exhibit of the tracks. A letter from F. N. Fisher, president of Galloway Coal Company, sent to Simpson on Memphis Park Commission letterhead and dated April 18, 1930, indicates that Simpson selected fourteen slabs of tracks directly from the No. 11 Mine tunnels to be shipped to him at the AMNH.[14] However, he was not getting the slabs for free; he had made an agreement with Galloway Coal Company to send some mammal fossils in exchange for them. He must have been slow to deal with the slabs, because they were eventually sent to the Memphis museum, and the tone of Fisher's letter expressed frustration that the exchange had not yet occurred. In the end, only two of the fourteen selected specimens were actually shipped to the AMNH, neither of which have been located.[15] Two of the remaining twelve slabs were put on display for many years at the Memphis museum and have recently been located.[16] The other ten appear to have been lost.

Although Simpson published no research on the Alabama tracks, he did write a brief unpublished report on what he thought about them, apparently in response to a request from W. Frank Cobb Sr., who may have been the one who originally invited Simpson to come to Alabama. Simpson's description of the animals and the time in which they lived is not that far off from modern interpretations, except for the revision of the age of the tracks from 250 to 313 million years. Some of his comments were part of the press release about his visit.[17]

The discovery of Paleozoic footprints near Carbon Hill is one of the great events in the history of Alabama paleontology. Both Jones and Simpson knew that the discovery was significant and had to be properly documented, even if few of their contemporaries saw the value of trace fossils for understanding the anatomy of the extremely ancient creatures that made them. This is what makes Museum Paper 9 one of the great legacies of Alabama natural history studies. When tracks were rediscovered in Walker County seventy years later by scientifically inclined amateurs, Museum Paper 9 became known to people who could understand the value of the work and the context of the tracks in the geological history of Alabama. This outcome was not inevitable. During the seven decades after the publication of Museum Paper 9, miners and others undoubtedly found many fossil tracks in numerous active mines that came and went over the years. Lacefield and Relihan (2005) described how, in the early 1990s, small groups of people associated with the University of North Alabama collected many slabs of high-quality tracks from the mine near Kansas, yet little research came out of those samples and most of that material was lost to science. The Birmingham Paleontological Society brought something different to the table: the level of enthusiasm and knowledge that was needed to bring the ancient footprints back to the forefront of paleontological research in Alabama. Let us now turn to those amateurs and what they found when they first visited the Union Chapel Mine.

8

Treasure Hunts

The first Birmingham Paleontological Society (BPS) visits to the Union Chapel Mine led to several important discoveries that highlighted the extraordinary nature of the fossil site. Here we recount a few of the finds from this time. Some of the described specimens were donated by the collectors to museums or institutions in Alabama, including the Alabama Museum of Natural History (ALMNH), the McWane Science Center (MSC), the Anniston Museum of Natural History, and the Geological Survey of Alabama. Appendix 2 provides information on the institutional or other locations of the specimens described in this and other chapters.

The first trip took place on January 23, 2000. This particular trip was special because the mine was a new site for the group and had not been visited by other amateur collector groups. Ashley Allen had scouted it only the month before this trip. When people arrived they spread out to look over a large area covered with numerous piles of large and small rocks, the so-called spoil piles from the mining operation (fig. 8.1). These piles included the characteristic thinly bedded gray shale that formed from sediment deposited when the area was an extensive mudflat. (Note the numerous flat rocks at lower left in fig. 8.1.) It was in these rocks that tracks and other trace fossils would be found. Mixed with these were many rocks from higher layers containing plant fossils, as well as rocks from sandstone layers that formed when the area was a marine environment.

At the time of this visit, the rock piles were treacherously high in places, and searching for tracks involved balancing oneself precariously

FIGURE 8.1 Spoil piles of gray shale at the Union Chapel Mine, January 23, 2000, the day of the first BPS visit. Part of the highwall is visible on the left.

FIGURE 8.2 Two spectacular footprints (manus and pes, nearly overlapping; UCM 124) of a large animal discovered by Jim Lacefield during the January 23, 2000, BPS field trip. The specimen is now recognized as *Attenosaurus subulensis*, the tracks of a protoreptile. Both prints are fairly deep undertracks. The white label has a long dimension of 6.7 cm (2.6 in).

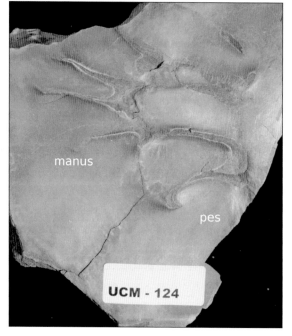

FIGURE 8.3 Small amphibian trackway (UCM 285) found by Ashley Allen on the January 23, 2000, BPS field trip; later recognized as *Matthewichnus caudifer*, the tracks of a temnospondyl amphibian (chapter 15). The tail drag indicates that the trackway was made at the primary surface.

on steep slopes. Although the lighting was unfavorable for seeing tracks, people still found interesting specimens. One of the most exciting discoveries of the day was made by *Lost Worlds* author Jim Lacefield, who found the rather large tracks shown in figure 8.2. As a past collector of tracks from other Alabama coal mines, he had never seen such big fossil footprints before. He was aware that some tetrapods reached large sizes in the Pennsylvanian Period, but this was the first time he had encountered evidence for them in Alabama (J. Lacefield, pers. comm. 2000).

On the same day, while a group of collectors searched another area, Ashley Allen turned over a small slab and found a beautiful trackway with small, well-defined footprints (fig. 8.3) later identified as *Matthewichnus caudifer*, thought to have been produced by a temnospondyl amphibian (chapters 4 and 15). Surprisingly, this specimen is one of only a very few of this type ever found at the site. The trackway is unusual in that it shows a clear meandering tail-drag mark, an indication

FIGURE 8.4 *Cincosaurus cobbi* (UCM 18) with intermediate-sized tracks, found by Steve Minkin. The specimen is lightly coated with a sealant.

that the footprints are surface tracks, not undertracks, as the majority of specimens would turn out to be.

The following weeks and months yielded ever more interesting discoveries. A core group of BPS members started to visit the mine more frequently. Steve Minkin found the exceptionally large *Cincosaurus cobbi* trackway in figure 8.4, one of the best of that type ever found at the UCM. It is currently on display at the ALMNH.

Many scientifically-important specimens from the Minkin Site have been discovered by Dr. T. Prescott Atkinson, Professor and Director of Allergy and Immunology, Department of Pediatrics, University of Alabama at Birmingham. Atkinson grew up in Alabama and hunted fossils as a child. When he was seventeen, he made a significant paleontological discovery: a dinosaur egg, the only fossil of its type that has been collected east of the Mississippi River. The egg was found in the Mooreville Chalk, part of the rock unit called the Selma Group. Prior to Atkinson's discovery, not even fragments of eggs had been found in these layers. Atkinson attributed the find to a "miraculous event" in which an egg was washed out to sea in a storm. Years later it was discovered that the egg contained the well-preserved embryo of a hadrosaur (a large vegetarian dinosaur).

After a several-year hiatus from collecting because of college and medical school (Emory University), Atkinson returned to amateur paleontology in the late '90s. He had joined the BPS only six months before Ashley Allen's presentation on the UCM trackways (chapter 1) and recalls what he thought at the time: "When [Allen] brought the tracks, I remember how beautiful they were. The shale that comes out of the site has kind of a polished appearance, really very fine grained, hard almost like china, really fine stuff" (P. Atkinson, pers. comm. 2009). Atkinson was excited about the possibility of collecting such fossils. He was not at the mine for the first BPS visit on January 23, 2000, but participated in a second visit with Jim Lacefield. Like Lacefield, Atkinson is an Indiana Jones "it belongs in a museum" collector, donating important specimens to institutions where they can be made available for study by professional researchers.

Two of Atkinson's biggest finds at the UCM were of fossil insect wings (fig. 8.5*A*) and the huge single track he named "Frogzilla" (fig. 8.5*B*). He found the wings in July 2000 in the middle of a dirt road that went down to the base of the highwall. The rock with the wings had apparently fallen from an adjacent embankment (Atkinson 2005). As is typical, Atkinson found both the impression and the counterimpression of the wings. From the way the two wings looked, he thought they were only half of an insect that originally had four wings. He looked hard to find the other half (as well as an impression of the body) but was unsuccessful. As we describe in chapter 16, virtually all the wings discovered at the UCM (Atkinson later discovered more) are of previously unknown species.

Atkinson further recalls the day he found Frogzilla: "I was in an overgrown area where there were a few slabs lying around. I split one open. It was not a big rock but it [the track] was positioned in the right place. The track was nicely placed right in the middle of it, a deep undertrack, possibly distorted, but probably the actual track was not much

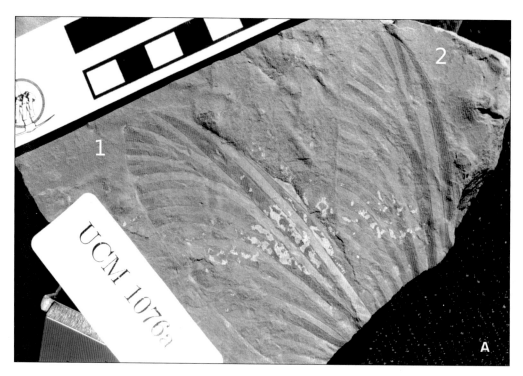

FIGURE 8.5 Important specimens found by T. Prescott Atkinson: (*A*) fossil wings of a huge insect (*Anniedarwinia alabamaensis*; UCM 1076a); the individual wings are labeled 1 and 2; (*B*) a very large *Attenosaurus subulensis* footprint (UCM 1621, dubbed "Frogzilla"), by far the largest yet discovered at the Minkin Site. The specimen is likely a deep undertrack. The middle toe impression is 17 cm (6.7 in) in length.

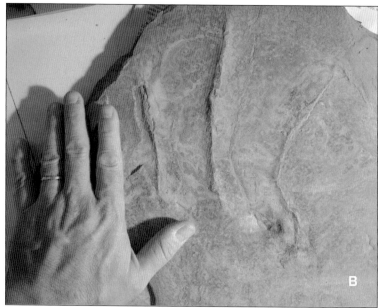

FIGURE 8.6 Holotype specimen (UCM 1142) of new ichnospecies *Nanopus reidiae*, thought to be made by a temnospondyl amphibian. Note the four digits in the small manus and five in the larger pes. Found by T. Prescott Atkinson.

larger than found. I couldn't believe my eyes it was so big. Probably it is not that far off from the size of the animal" (P. Atkinson, pers. comm. 2009). After more than ten years, no other comparable-sized specimen of this type has yet been found.

The insect wings and Frogzilla are only two of many important fossils Atkinson found over the first year of UCM searches. He also discovered UCM specimen 1141/1142 (fig. 8.6), which became the holotype (defining example) of what Haubold et al. (2005a) would name *Nanopus reidiae*, thought to be the tracks of a second type of small temnospondyl amphibian (chapter 15). Neither *N. reidiae* nor *Matthewichnus*

FIGURE 8.7 Specimen UCM 1075, a large slab that is cracked in many places but that shows the strongest evidence for group behavior of coal-age tetrapods: *top*, the actual specimen; *bottom*, processed version using unsharp-masking to enhance the tracks, identified as *Cincosaurus cobbi* by H. Haubold (chapter 15). Found by T. Prescott Atkinson.

caudifer had been found at Carbon Hill in the 1920s. Atkinson also discovered UCM 1075 (fig. 8.7, top), a larger slab covered with *Cinco-saurus cobbi* tracks of several animals that appear to have been moving together. If this is what the specimen actually shows, then it provides some of the earliest evidence of group behavior in reptiles. The trackway

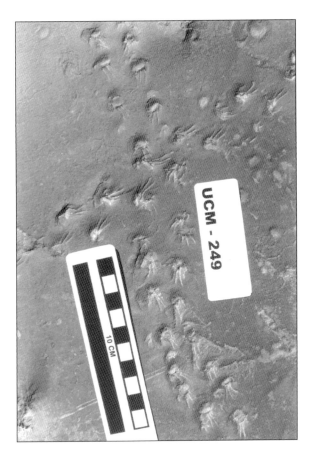

FIGURE 8.8 Tracks made by a small temnospondyl amphibian (*Nanopus reidiae*) on a large slab found by Jay Tucker.

horizon could not be retrieved as a single slab but was reconstructed like a puzzle from individual pieces. The same specimen is shown at the bottom of figure 8.7, but we've used unsharp-masking to enhance the footprints.[1] This processed image dramatically reveals the clear multiple trackways and their similar direction of movement.

Jay Tucker found the spectacular footprints shown in figure 8.8. After splitting a large slab, he found two extremely well-defined trackways of a small amphibian (*Nanopus reidiae*). The close-up shows mostly pes traces with conspicuous oval pads. Manus traces are smaller and mostly diffuse. Over a hundred footprints, including both manus and pes, are found in this beautiful specimen. The two trackways are so similar in width and footprint size that they might have been made by the same animal.

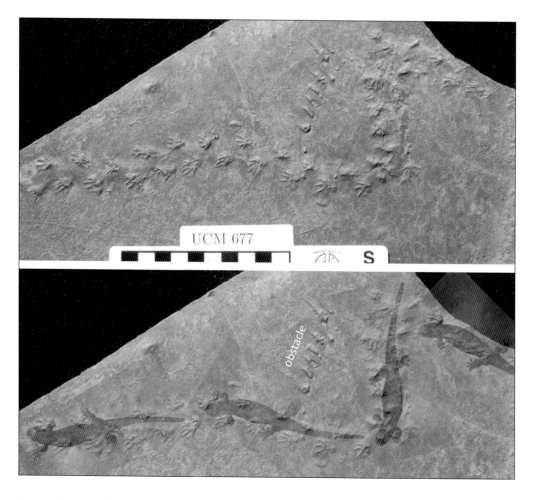

UCM 677

obstacle

FIGURE 8.9 *Top*, tracks (*Nanopus reidiae*, in positive hyporelief) made by a small amphibian walking around an obstacle; *bottom*, a phantom trackmaker, approximately 10 cm long, is imagined as it walks around the obstacle and leaves its footprints. Specimen found by Ron Buta (chapter 15).

Most of the vertebrate tracks found during the early BPS visits were of small animals simply walking across a surface. That is, the trackmakers do not appear to have been doing anything other than walking (or meandering) in a given direction. But fossil trackways can sometimes reveal more. Coauthor Ron Buta found the specimen shown in figure 8.9 (top). A small amphibian (maker of *Nanopus reidiae*) appears to have walked around an obstacle. The obstacle is in the form of six or seven short, parallel striations and does not appear to be plant matter. It could be a partially buried horseshoe crab (Martin and Pyenson 2005). The trace is an elevated counterimpression, meaning we are seeing the

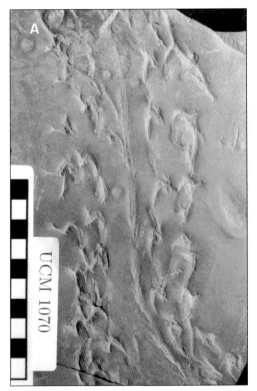
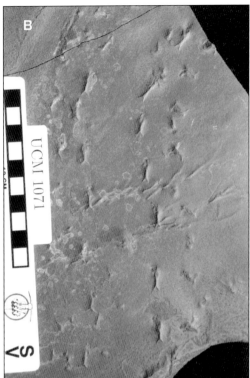

trackway from the underside. When the animal came upon the obstacle, it took a sharp right, then a sharp left, and walked on (fig. 8.9, bottom). The specimen was found only after splitting a large slab, and it reveals a specific type of behavior—obstacle avoidance. Only a few other examples of this type have been found at the site.[2]

Although collectors focused mostly on vertebrate traces initially, they also found high-quality invertebrate traces during these early visits. Atkinson found the specimens shown in figure 8.10*A* and *B*, in which horseshoe crab trackways (*Kouphichnium*), both surface tracks and undertracks, are extremely well preserved. Atkinson also found some of the first examples of the unusual repetitive trace shown in figure 8.11, which at first sight looks like a series of body impressions. Atkinson suspected that these were "jumping" traces made by an unknown animal. An alternative interpretation proposed at the time was that these traces

FIGURE 8.10 These two beautiful trackways (*Kouphichnium*; chapter 16) were made by the same animal (a small horseshoe crab) at the same time but formed in the mud at different horizons. The tracks are on opposite sides of the same slab: (*A*) UCM 1070; (*B*) UCM 1071. *B* is an undertrack relative to *A* and shows little more than two double rows of parallel dash- and *Y*-shaped features, far less than what UCM 1070 shows. Found by T. Prescott Atkinson.

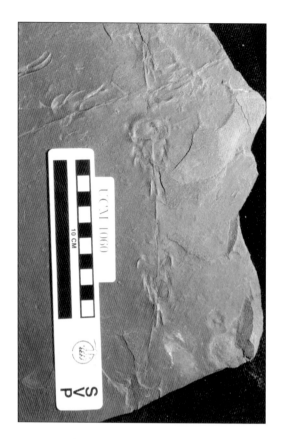

FIGURE 8.11 Jumping traces (UCM 1060) identified as *Tonganoxichnus robledoensis* and made by a wingless insect (chapter 16). Found by T. Prescott Atkinson.

were caused by burrowing horseshoe crabs moving in and out of the sediment. The best interpretation of these unusual traces advanced so far is that they were made by a wingless, jumping insect (chapter 16).

Related to these traces, Bruce Relihan, former curator of horticulture at the Birmingham Zoo, found a large slab with multiple layers that, when separated like the pages of a book, revealed numerous trackways that at the time were thought to be *Kouphichnium* (fig. 8.12, top). Later studies showed that these traces were a form called *Stiaria*, attributable to the same wingless insect that produced Atkinson's jumping traces (chapter 16). Relihan also found the spectacular trackway shown at the bottom of figure 8.12, which consists of two parallel sets of small dots crossing a large slab. Later studies attributed this kind of trace (called *Diplichnites*) to a primitive millipede.

UCM 953

FIGURE 8.12 *Top*, dense horizon of insect trackways (ichnogenus *Stiaria*, UCM 485k), one of five horizons of such traces in a single slab; *bottom*, beautiful trackway likely made by a millipede and identified as *Diplichnites gouldi* (chapter 16). Both found by Bruce Relihan.

These and other discoveries show the exceptional nature of the UCM fossil assemblage. The ichnofauna is more diverse than at any other mine in the Black Warrior Basin. Not only are the specimens abundant and of very high quality, but they include remarkable examples illustrating behaviors, like group or herd behavior, that are rarely preserved, as well as rare arthropod body fossils. The plant fossils from the Minkin Site are spectacular as well. Chapter 17 includes photos of well-defined bark impressions of *Lepidodendron*, *Lepidophloios*, and *Sigillaria*, three abundant arborescent lycopods from the period. For BPS members who explored the site in those early days, the UCM became more than just another field trip site. It was inspiring to go there and find the footprints of animals that lived long before the first dinosaurs. People like Steve Minkin were exhilarated by the experience.[3] These were ancient footprints that ordinary people could find and collect, unlike dinosaur footprints, which generally belong in the domain of professional paleontologists, and in any event are rare in the southeastern United States. Also, ordinary people could contribute to the rescue of scientifically important specimens from a site that, at the time, was destined for destruction. The fear of impending reclamation drove a great deal of the collecting activity during the first year and set the stage for the BPS's attempt to preserve the site.

A major deficiency of collecting trace and plant fossils from spoil piles is the loss of geologic context. While spoil piles will not necessarily be far from where the rocks were extracted, the positions from which scattered fallen rocks originated are lost. More than 25 m (82 ft) of rock layers were exposed at the Minkin Site during the earlier visits, but it was not obvious to any of the collectors which layers the tracks and plants came from. To figure this out, as we shall see in the next chapter, a more scientific approach was needed.

9

Anatomy of a Highwall

There is an important concept in geology known as stratigraphic (or geologic) context. As we described in chapter 2, the nature of sedimentary rock implies that the ordering of the layers is a time sequence, with younger layers in the geological column overlying older layers (i.e., Nicolas Steno's Law of Superposition). If a sedimentary rock formation includes specific fossils in specific layers, then knowing in which layers these fossils are found pinpoints the exact position of the fossils in the time sequence. Observing the characteristics of those layers, moreover, may shed important light on the environmental conditions that existed when the animals that left the fossils were alive.

In the case of the No. 11 Mine of the Galloway Coal Company, there was never any ambiguity about the source of the trackways. Unlike surface mining, underground coal mining does not involve significant removal of overburden. Miners follow a seam and extract the coal layer directly from the subsurface. Upon extraction, some of the roof strata above the coal seam commonly collapse into the mine cavity. When this was done for the Jagger coal seam at the No. 11 Mine, the shale layers within a meter (a few feet) above the top of the seam were immediately recognized as the source of the tracks. Indeed, tracks were seen in the ceiling of some tunnels, where the roof shale had collapsed.

In the case of the trace fossils from the Minkin Site, stratigraphic context was mostly lost because the fossils could not be collected in situ (in place). The surface-mining operation removed tons of rocky overburden in order to reach the long-buried coal reserves, and the fossils appeared in some of the piles of displaced rock. To an average collector who was not there when the spoil piles accumulated, it would not be obvious which layers of rock the trace fossils came from. However, if the layers containing the trace fossils had some distinctive characteristic that could be recognized even when the rocks were not in stratigraphic context, then it would be possible to identify those layers directly in the highwall created by the mining operation. For example, imagine that the track-bearing rocks found in the spoil piles were always bright red, while all the other rocks lying around were dark gray. Then all you would have to do is look at the highwall and find a bright red layer against a background of gray rock. This would tell you exactly where the tracks came from. Of course, nature is rarely that convenient. There may very well be some distinctive characteristics of the track-bearing layers in the side view of a highwall, but it may take a professional sedimentary rock specialist to recognize them.

The BPS was very fortunate to have Dr. Jack C. Pashin involved in the stratigraphic study of the Minkin Site fossils. Pashin was the head of the energy investigations program of the Geological Survey of Alabama (GSA) at the time of the Union Chapel discoveries.[1] His research interests are coal-bed methane, coal geology, and petroleum geology. He received a bachelor's degree from Bradley University in Peoria, Illinois, and a PhD in geology from the University of Kentucky. His group at the GSA has studied all aspects of the state's energy resources: coal, oil, and gas, and more recently, carbon sequestration and climate change technology. His PhD dissertation was on the sedimentology at the Devonian-Mississippian boundary in the Appalachian region. He is also a past president of the Alabama Geological Society and a past chair of the Coal Geology Division of the Geological Society of America.

Pashin first learned about the UCM discoveries at an Alabama Coal Association banquet, where he met New Acton Coal Mining Company owner Dolores Reid.[2] Recall from chapter 1 that Reid is the grandmother of Jessie Burton, the student in Ashley Allen's class who first brought attention to her coal mines. She told Pashin that collectors were finding interesting fossils at one of her mines, the UCM, and shortly thereafter, he learned that many of the fossils were trackways. As he found out more about it, he thought someone needed to visit the site to try to determine where all the trackways were coming from. "That's ultimately how I got involved," he said (J. Pashin, pers. comm. 2010).

FIGURE 9.1 The highwall at the Minkin Site, an unstable rock face. At the time of the first Union Chapel trackway discoveries, this cliff face was about 25 m (82 ft) high.

For Pashin, determining where the tracks at the UCM came from meant getting as close a look as possible at the layers in the highwall and seeing how they were assembled and sequenced in terms of rock type and other details. The ordering of exposed rock layers at a given sedimentary rock site is called a succession, and measuring those layers as a function of height above ground level (i.e., making a stratigraphic section) provides an important record of the general history of the site over the period represented by the section. Different kinds of rocks, such as shale, sandstone, siltstone, mudstone, coal, or conglomerate, can be recognized from their color or texture in the highwall. In addition, any cyclicity in the rock layers, such as repetitive thickening and thinning of beds, evident bioturbation (visible evidence of burrowing), or even whole standing tree trunks, needs to be recognized since such features can point to specific environmental conditions. Because virtually every surface coal mine in Alabama has an exposed highwall, professional paleontologists working at such sites must measure the section as well as collect scattered fossils from spoil piles.

Naturally, measuring the section at the Minkin Site was going to be challenging: as shown in figure 9.1, the highwall there is a treacherous-looking, almost vertical cliff that was more than 25 m (82 ft) high after mining operations ended. At the same time, because of the way it was made, it was not stable. (Recall in chapter 1 the story Dolores Reid told about a rock slide from the highwall that occurred the day Ashley Allen's class was visiting.) Given this inherent danger, how could the highwall be studied in the detail needed to determine where the tracks came from? Essentially, you put on your safety gear, find the safest-looking place to walk along the edge of the highwall, and scale it as far as is possible. This is what Pashin did on August 8, 2000.

FACE TO ROCK FACE

From regional studies, both his own and those previously published by other scientists, Pashin knew before he looked at the Minkin Site highwall that trackways probably came from somewhere between the two coal beds that had been mined there. This is exactly what he found by direct observation at the site. He climbed up as high as he could so he could look at each layer in place and collect samples where possible.[3] He summarized the results of his study in an excellent article published in the Alabama Paleontological Society's 2005 monograph on the UCM fossils (Pashin 2005). In this article he graphically displayed his analysis of the rock layers in the highwall in a stratigraphic section. The section used horizontal boxes with different shadings and thicknesses to represent the different rock layers at different vertical positions.

Figure 9.2*A* shows what Pashin called a "generalized stratigraphic section" of the entire Pottsville Formation in the Black Warrior Basin of Alabama. The diagram shows the types of rock at different levels of the formation, ranging from coal beds to interlayered (interbedded) shale and sandstone, to thick sandstone layers, and finally to conglomerate layers, rocks containing rounded pebbles (clasts) cemented together with finer material such as sand. More than thirty coal beds have been identified in the Pottsville, which is nearly 2 km (1.2 mi) thick in Walker County. These beds are not randomly distributed but cluster into groups called coal zones, whose names are indicated in figure 9.2*A*. The coal mined at the UCM was part of the Mary Lee coal zone, which is an interval of rock 60 m (200 ft) thick that sits on a thick layer of sandstone. The Mary Lee zone is bounded at the top by a marine zone

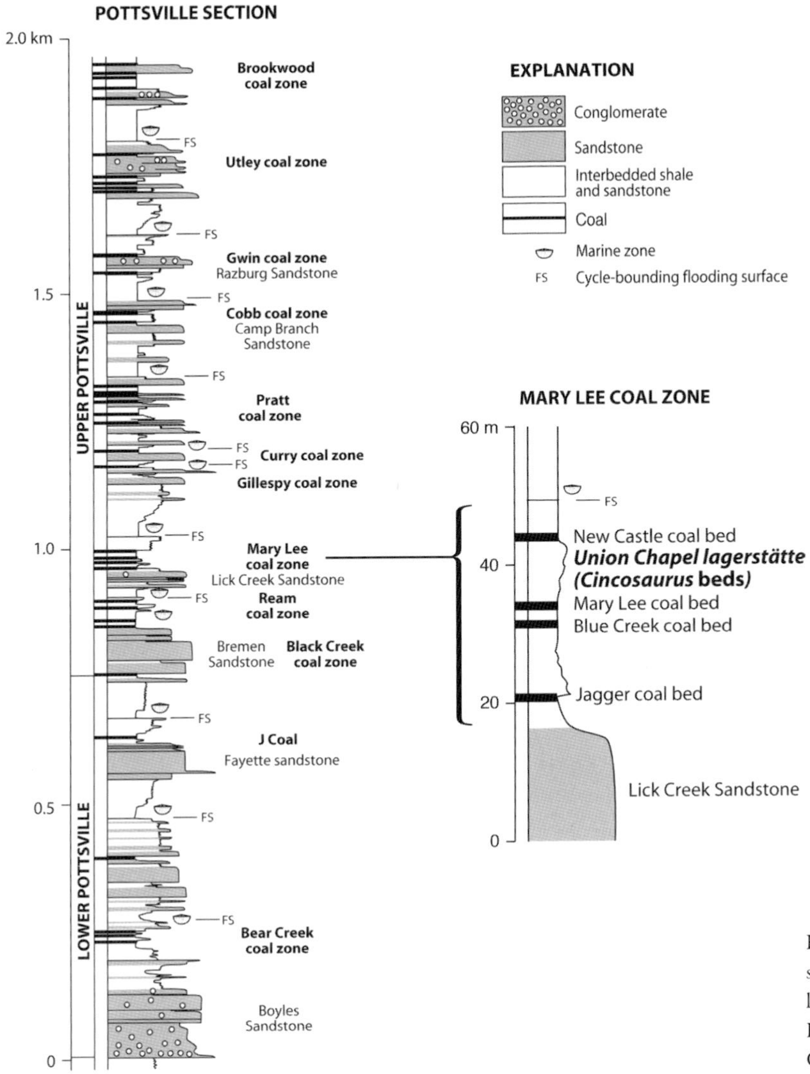
POTTSVILLE SECTION
2.0 km
1.5
1.0
0.5
0
UPPER POTTSVILLE
LOWER POTTSVILLE
Brookwood coal zone
FS
Utley coal zone
FS
Gwin coal zone
Razburg Sandstone
FS
Cobb coal zone
Camp Branch Sandstone
FS
Pratt coal zone
FS
Curry coal zone
FS
Gillespy coal zone
FS
Mary Lee coal zone
Lick Creek Sandstone
FS
Ream coal zone
Bremen Sandstone
Black Creek coal zone
FS
J Coal
Fayette sandstone
FS
FS
Bear Creek coal zone
Boyles Sandstone
EXPLANATION
Conglomerate
Sandstone
Interbedded shale and sandstone
Coal
Marine zone
FS Cycle-bounding flooding surface
MARY LEE COAL ZONE
60 m
40
20
0
FS
New Castle coal bed
Union Chapel lagerstätte (Cincosaurus beds)
Mary Lee coal bed
Blue Creek coal bed
Jagger coal bed
Lick Creek Sandstone

Figure 9.2*A* Stratigraphic section (map of rock layers) of the Pottsville Formation in Walker County.

whose rock layers formed when sea level was higher (Pashin 2005). The Mary Lee zone includes four named coal beds (from top to bottom): the New Castle, the Mary Lee, the Blue Creek, and the Jagger. Only one of these beds remained exposed in the highwall when mining operations ended at the Minkin Site. At the time of his study, Pashin identified the exposed bed as the New Castle and inferred that the Mary Lee bed

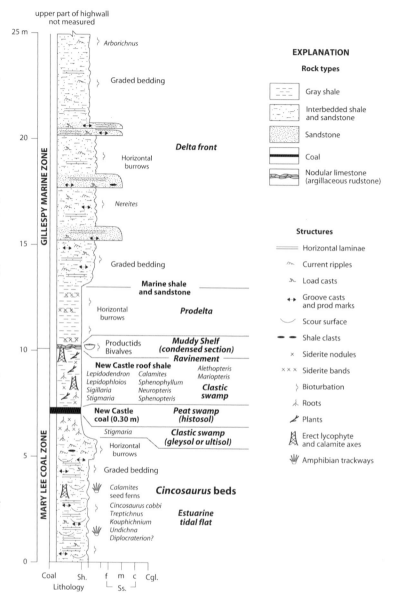

FIGURE 9.2B Stratigraphic section (map of rock layers) of the Union Chapel Mine highwall.

Schematics 9.2A and 9.2B are sophisticated recordings of the sedimentary rock layers and how they "stack" up. Coal beds are indicated by dark horizontal lines and are not randomly distributed, but cluster into coal zones. Shale, sandstone, and conglomerate layers have different symbols. In figure 9.2A, a flooding surface FS signals a rise in sea level and is usually followed (in the time sequence) by rocks deposited when the area was a marine zone. In figure 9.2B, much finer detail in the rock layers shows where various fossils, of both tracks and plants, likely came from. The geologic terms, such as load casts, groove casts, prod marks, and so forth, describe different rock characteristics having to do with sediment properties and sequence of deposition, and of water flow behavior.

was likely only a few feet below ground level (fig. 9.2B). He concluded that the vertebrate trackways were coming from layers of shale between these two coal beds. The Blue Creek and Jagger seams were thought to be absent from the Minkin Site. The tracks found near Carbon Hill between the Jagger and Blue Creek beds were from an older stratigraphic interval, according to this interpretation. However, coal beds can be difficult to follow over large areas. They may be discontinuous or disappear altogether, so that it may be confusing as to which beds in the Mary Lee coal zone are actually being mined in a given location. From a more recent study of the highwall at a different mine, the Crescent Valley Mine near Carbon Hill, Pashin has concluded that tracks at the Minkin Site and those found by Aldrich and Jones at the No. 11 Mine are probably from the *same stratigraphic interval*, and hence are of the same age (Buta et al. 2013). This is discussed further in chapter 19.

Figure 9.2B provides more detail of the rock layers at the Minkin Site. Various symbols indicate what kinds of fossils came from which layers. Particularly interesting is how layers rich in plant fossils are not necessarily rich in tracks. The "roof shale" of the exposed coal bed (that is, the shale just above the coal bed labeled New Castle in fig. 9.2B) is where many of the plant fossils came from. Plant fossils are found in the *Cincosaurus* beds also, but not as abundantly or in such great variety. The layers of rock forming the roof shale of the coal bed were almost certainly deposited tens of thousands of years later than the footprint layers below the coal bed. Apparently, the area was first a dense tropical wetland forest (leading to the formation of what was originally identified as the Mary Lee coal bed), then it was a tidal mudflat (leading to the deposition of the *Cincosaurus* beds), and then a tropical wetland again (preserved as what is identified as the New Castle bed). Tracks occur only in rock layers laid down on mudflats. A forest would have grown nearby and was in all likelihood where the animals actually lived. This partly explains why all coal mines in Alabama have plant fossils but not necessarily tracks. If the tidal-flat deposits are not preserved, there is less chance of finding tracks.

By examining the rock layers in the wall and comparing them with the track-bearing rocks in the spoil piles, Pashin deduced that the *Cincosaurus* beds, the layers containing the vertebrate footprints, came from 2 to 3.5 m (6 to 12 ft) above ground level, or 3.5 to 5.5 m (12 to 18 ft) below the exposed coal bed. Since the time of Pashin's measurements, the lower part of the highwall has been filled in and the track-bearing layers are no longer above ground.

Figure 9.2*A* shows that the Mary Lee coal zone is in the middle of the Pottsville, and that many other coal beds lie above this particular zone. Pashin's analysis, however, showed that the top of the highwall at the Minkin Site extends only into marine layers of the Gillespy coal zone, the one immediately above the Mary Lee. What happened to the coal beds from these higher coal zones? The answer is that in the area of the UCM, these coal zones in the Upper Pottsville eroded away long ago. If we could have waited long enough, erosion would have exposed the Minkin Site trackways naturally, without a mining operation. But natural erosion can be a very slow process. To reach the track-bearing layers naturally could have taken hundreds of thousands of additional years. In north-central Alabama, nature has brought the land down so far that some of the most economically productive coal layers are near the surface, and coal mining is practical.

Tidal Rhythmites

Pashin identified the rock layers in the highwall that were the source of the trackways by comparing collected track-bearing slabs with what he saw in the highwall. He noticed that the *Cincosaurus* beds consisted of stacked laminae (thin flat layers) ranging in thickness from less than a millimeter to several centimeters. The trackways were preserved on the surfaces of some of the laminae. What was the nature of these laminae? Most people are familiar with storm deposits, event layers formed by the erosion and redeposition of sediment by major storms. Go out in the street after an intense rain and you will find sediment (pebbles, sand, dirt) moved by runoff from wherever it was yesterday to a new, lower position. But most of the very thin layers making up the *Cincosaurus* beds are probably not storm deposits. They show a *characteristic periodicity* that strongly suggests the influence of tides (Pashin 2005). A tide is a periodic rising and falling of water in a coastal environment that occurs because of the gravitational influence mainly of the moon and sun. Differential gravitational forces raise water levels on opposite sides of the Earth, producing what is called a tidal bulge. The tides occur because the Earth and the moon are large compared to their average separation, which allows the moon's gravitational pull on the side of the Earth facing it to be significantly larger than its pull on the opposite side of the Earth.

The evidence for tidal influence at the Minkin Site is shown in figure 9.3. These are side views of "pinstripe-bedded" mudstone. The stripes are caused by sediment grading: siltstone or sandstone in the light, thin stripes, and shale or mudstone in the dark, thicker areas. Siltstone and sandstone are made of heavier particles than mud, and each light-colored stripe is the bottom of a graded bed. This pinstripe-bedded mudstone tells us that sedimentation in the *Cincosaurus* beds was periodic (occurred at a regular frequency, or "rhythm") and has all the hallmarks of tidal influence, with each graded bed representing a single event. To make this process work, you need a major source of sediment, such as a river flowing from nearby mountains, meeting an area subject to daily tides. As we noted in chapter 5, this would occur in an estuary, a place where a river flows into a sea. At high tide, the sedimentation process would be in full swing. Heavier particles of silt or sand would fall to the bottom more quickly than lighter particles, leaving a light-colored pinstripe at the bottom of the newly forming bed. Progressively light-er-weight and darker particles would settle out as currents diminished until low tide and slack water, which would halt the sedimentation

FIGURE 9.3
"Pinstripe-bedded" mudstone in the highwall of the Minkin Site. This is the kind of rock that characterizes the *Cincosaurus* beds. Each graded bed (light to dark) records a single cycle in a tidal sequence. The lighter stripes represent the heavier particles that settled at the beginning of the cycle. *A* shows a gradual decrease then increase in bed thickness, while *B* shows paired thick and thin beds.

process temporarily. At this time, the area would be an exposed mudflat where animals could wander out looking for food and leave their trails of footprints and other traces. Then when the tide came in again, the cycle would repeat: the river would bring fresh sediment to the water that would settle in the same manner as in the previous cycle: heavier particles first, then lighter particles, leaving stacked pinstripe-bedded laminae that eventually became solid rock.

Sedimentologists can easily recognize the tidal influence in the rock record because the characteristics they see in these rocks are identical to what they find in tidal settings all over the world today. According to Pashin, sedimentation patterns in tidal settings are unique. Tides can lead to the graded beds shown in figure 9.3, and even the subtleties of those beds, such as the systematic change in bed thickness seen in figure 9.3*A* and the paired thinner and thicker beds seen in figure 9.3*B*, can be explained in terms of varying tide strength resulting from the changing phase of the moon and its location in its orbit. Because the rotation of the Earth, the orbital motion of the Earth, and the orbital motion of the moon are all periodic, all variations in tide strength will be periodic to some degree. At the tropical location of the Minkin Site, we would have expected two tides a day, but these could have been of unequal strength depending on where the moon was, with the lesser tide producing a narrower graded bed because of lower sedimentation (Kvale 2006). In figure 9.3*B*, each pair of thick and thin beds represents one day's tides at a time when the two tides were significantly unequal. Figure 9.3*A* shows a progressive thinning of the beds from lower layers to higher layers, which can be attributed mostly to the difference between spring tides (where the Earth, moon, and sun are roughly aligned, making tides larger and sedimentation higher) and neap tides (where the moon and the sun are about 90° apart, making tides weaker and sedimentation lower). This effect would occur with a period of about fourteen to fifteen days, or half of a lunar phase cycle (Pashin 2005). Tidal deposits at the Minkin Site are less regular, suggesting that sedimentation was strongly influenced by something in addition to the moon and the sun, perhaps the varying input of water from rivers. The site may have been farther up a river channel and not right out in the main part of the estuary.

The idea that the Minkin Site was a tidally influenced, freshwater estuary is supported by the properties of the coal that was mined there. This coal has an extremely low sulfur content, which implies that freshwater inundated the swamps that became the coal seam. Pashin concluded that "the *Cincosaurus* beds can be interpreted as an estuarine mudflat in which tidal currents primarily moved freshwater about" (Pashin 2005).

Modern-day Alabama has a place not very different from the Westphalian A estuary whose former existence is preserved at the Minkin Site. It is called the Mobile delta. Here, freshwater enters the large estuary we call Mobile Bay. Lush foliage on the delta demonstrates the area's warm, wet climate. Rainfall has been more than 150 cm (60 in) per year on average in historical times. This is almost enough to qualify southern Alabama as a rain forest. To be counted as a rain forest, by one commonly accepted definition, a wooded area must receive at least 173 cm (68 in) of rain per year. An even better analogue for the Mary Lee environmental setting might be the tropical tide-dominated deltas of Indonesia, where peat is accumulating in major rain forests (Pashin 2005). The *Cincosaurus* beds are sandwiched between two coal seams, and coal is nothing but the remains of ancient rain forests. We do not know how much rain fell per year while the *Cincosaurus* beds were forming, but it must have been substantial.

How would the myriad small footprints have been made and preserved at the Minkin Site? Pashin concluded that all the trackways at the Minkin Site would have been made at the tops of the graded beds (that is, just below where a pinstripe would have developed; fig. 9.3) (J. Pashin, pers. comm. 2010). This represents a time of low water and is where the primary tracks would have formed. However, undertracks could have been preserved at the bases of the same graded beds or even at deeper levels.

We have already noted in chapter 5 that fossilized mud cracks are very rare at the Minkin Site. This is the chief evidence that much of the exposed mud never had a chance to dry completely at low tide, and it explains why most of the primary surface tracks that have been collected at the Minkin Site are not well defined. The mud's surface was in general too saturated with water to preserve sharp impressions. Mud cracks are not rare; they are common in environments where mud has a chance to dry.

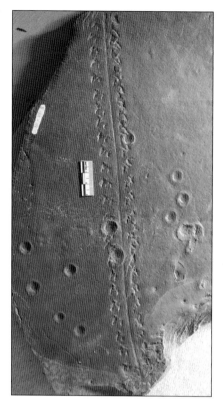

FIGURE 9.4 *Top Left,* This slab (UCM 1740), bearing a regular trackway of the ichnogenus *Stiaria* (chapter 16), includes about a dozen circular impressions, each about 1 cm in diameter, that were likely caused by gas bubbles rising from within the wet, organic-rich mud.

FIGURE 9.5 Modern raindrop impressions. *Top right,* Harrell Station, Alabama; *bottom right,* Leeds, Alabama. Black arrow indicates large raindrop imprint with splash marks.

The absence of mud cracks from shale at the Minkin Site suggests that small tetrapods made their way through soupy mud. Sometimes the mud was exposed to the air and sometimes it may have been submerged by a thin layer of water. The water could not have been very deep because most of the animals were small. Modern newts (salamanders) may sometimes be seen walking on the bottom of small ponds in up to about 15 cm (6 in) of water. However, the Minkin Site was at the edge of a major estuary when the *Cincosaurus* beds were deposited. Plenty of swimmers would have liked nothing better than to eat something the size of a salamander. Almost all the vertebrate trackways were made by animals that would have stayed close to shore to avoid becoming a meal.

GAS BUBBLES AND RAINDROPS

Some slabs in the *Cincosaurus* beds, both with and without trace or other fossils, have an abundance of small circular depressions or craters.

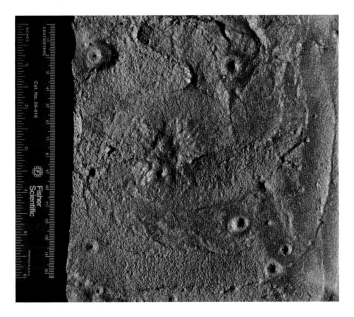

Figure 9.6. Modern gas bubble marks.

An example is shown in figure 9.4, where an invertebrate trackway (*Stiaria*; chapter 16) is flanked by several similar-sized depressions. These have been known since the early nineteenth century, and two competing explanations for them have been put forth (Rindsberg, pers. comm.). The first is that the marks were made by gas bubbles popping at the surface of the mud. These form in soft, wet mud and there is no need to suppose that the mud ever fully dried out. Alternatively, they could be raindrop impressions (fig. 9.5) made in wet but firm mud during light rainfall events. That is, they develop in mud that's already drying and are most easily preserved if the mud dries out completely afterward. This way it gets firm enough to survive being flooded with sediment and buried.

The gas-bubble interpretation can explain most of the circular depressions seen in Minkin Site rocks (Rindsberg 2005) because the depressions have characteristics that favor this interpretation. Gas-escape structures (fig. 9.6 shows modern ones) will have a characteristic size rather than a wide range of sizes, because gas bubbles have to be large enough to float upward through wet sediment. Also, two bubbles will not necessarily rise in the same place to leave overlapping craters, so the way the depressions are distributed on a slab is not random. The circular depressions on UCM 1740 (fig. 9.4) are excellent examples that satisfy

FIGURE 9.7 Probable raindrop impressions from the *Cincosaurus* beds, UCM 1024. Note the considerable overlap and size variation of the circular impressions.

these criteria, although other specimens can be ambiguous. Figure 9.7 shows what could be a genuine example of raindrop impressions from the Minkin Site.

Given that the *Cincosaurus* beds probably received quite a bit of rain each year, perhaps as much as a modern rain forest, and that there are nevertheless very few raindrop impressions preserved in the shale, we are drawn to one conclusion. The sediment that went to make up the rock that we call the *Cincosaurus* beds was either never exposed to the air before it became firm or solidified, or was exposed so briefly that it never had the chance to dry out. There ought to be, somewhere, rocks that were laid down higher up on the tidal flats. If we ever find them, we will recognize them because they will likely contain both indisputable raindrop impressions and mud cracks in abundance (unless they were colonized quickly by plants and converted to soils) (J. Pashin, pers. comm. 2013).

A diverse assemblage of plant fossils occurs on slabs of shale that also bear trackways, so we know these are plants that lived around the estuary when the *Cincosaurus* beds were being deposited. A few erect *Calamites* and seed ferns were observed in the *Cincosaurus* beds in the highwall, proving that some plants grew right out on the mudflat when the reptiles, amphibians, and various invertebrates whose trace fossils we have collected were walking around and living their lives (J. Pashin, pers. comm. 2013). Plant fossils are common in the *Cincosaurus* beds, but more were preserved higher in the section.

TIME

In the Pottsville Formation coal zones, trees are occasionally preserved standing upright. For each standing forest that has been preserved in the Pottsville Formation, how many years passed? We know from human history that preservation of standing forests is rare. Trees live and die all over the world for centuries before a mudflow, ash fall, lava flow, or other catastrophic event preserves a forest with trees still standing. How long would it take for multiple forests to be preserved this way, *in the same place*?

Pashin explained it this way: "The interval between two [major] coal beds probably represents 100,000 years . . . , and yet much of the sediment between those two coal beds looks like it has accumulated very very rapidly, which always begs the question, where is the time? It's certainly not in sediment that contains standing tree trunks." He suggests that the time is represented or "concentrated" in the fossil soil beneath a coal bed and in the coal bed itself (J. Pashin, pers. comm. 2010).

Sedimentation in most environments is not a steady process. A place where sand and mud have been deposited millimeter by millimeter, millennium after millennium, is rare and precious to the student of Earth's history. This is because natural processes like weather, the movements of rock deep in the Earth that cause most earthquakes, and volcanic eruptions do not proceed slowly and steadily. The day-to-day movement of sand on the beaches of Dauphin Island (southern Alabama) effects changes, but most of the change happens when a major hurricane makes landfall and cuts the island in half, or drops six feet of sand in the streets among what remains of the houses. Water flowing in the Black Warrior River carries a certain amount of suspended mud. But most movement of sediment by rivers occurs during major floods, when they burst out of their banks and rearrange the landscape in only hours or days. In 1993 heavy summer rains caused flooding that affected the entire Mississippi River Valley. Land was underwater that had been safe from the river as long as people could remember. It happened again in the spring of 2011, but not during the intervening eighteen years.

Many natural processes, not just weather, occur on this same kind of erratic schedule. Basically, the more intense an event is the less often it will happen. A hundred-year flood, with a statistical probability of occurring, on average, only once a century, is a flood so powerful that we expect to have to deal with one only once every hundred years. The recurrence interval for major earthquakes is a measure of how long we expect to wait before we experience another earthquake exceeding a given intensity in a particular location or fault zone. These estimates come from inputting real-world data into mathematical probability models. A qualitative understanding is sufficient in this case, especially because we do not have a lot of quantitative information about the rivers or storms that killed or preserved those fossil forests more than 300 million years ago.

So far we have been talking about only catastrophic events. A flood or hurricane produces an event deposit. This is a layer of gravel, sand, or mud that the flood or storm creates. We know a lot about event deposits from studying them in the modern world. Storm deposits vary in thickness from millimeters to tens of centimeters. Individual storm layers vary from place to place. Some are preserved in the sediment record and can still be seen hundreds of millions of years later. Many storm deposits are eroded by later powerful storms.[4] Others are homogenized, or mixed with underlying and overlying fair-weather deposits by burrowing organisms. This process can be completed in just a few years where burrowers are numerous and active. Because major storms erode sediment before depositing new sediment, fair-weather deposits tend to be scarce in the rock record wherever frequent storms leave their marks. Deposition is unsteady and preservation is even more so. Shale laminae at the Minkin Site, those bearing trackways as well as those that are smooth and unblemished, are like sheets torn out of a book. The book has been thrown away, and all we have left with which to puzzle out past events are a few isolated pages.

The *Cincosaurus* beds and associated coal seams at the Minkin Site were once buried by kilometers of overlying sediment. We know this because the coal has been compressed and heated. Pashin explained how

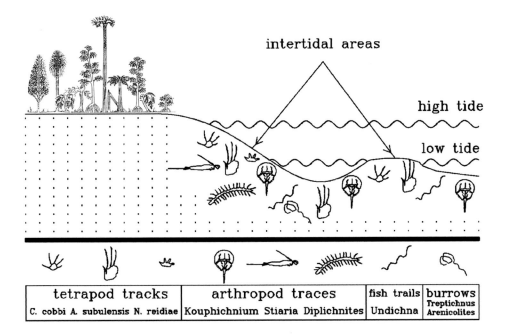

forest, swamp

intertidal areas

high tide

low tide

tetrapod tracks	arthropod traces	fish trails	burrows
C. cobbi A. subulensis N. reidiae	Kouphichnium Stiaria Diplichnites	Undichna	Treptichnus Arenicolites

we determine former depth of burial by "analyzing the chemistry and the physical properties of coal. When we look at Pennsylvanian rocks, we see . . . bituminous coal. It's been coalified quite strongly. And it gets less and less glassy as [you] go upward in the section. . . . When we reach the [surface where the layers are eroded away], we're missing the upper part of that glass series. . . . The sediment preserved above the unconformity has lignite in it, which is a very dull coal. It's not glassy at all. So we're missing the less glassy coals between the lignite and the very glassy coals preserved in the Pottsville Formation. And that's basically how we determine how much sediment is missing" (J. Pashin, pers. comm. 2010).

Figure 9.8 is a schematic summary of the results of Pashin's analysis. A tidally influenced, freshwater area lies near a swamp forest with all the characteristic plants of the period, including the towering *Lepidodendron* scale trees, tree and seed ferns, and horsetails (*Calamites*). During low tide, an expanse near this swamp forest is exposed, allowing a variety of tetrapods and arthropods to leave their footprints in the wet mud.

FIGURE 9.8 Schematic diagram of a tidally influenced area near a swamp forest, with symbols indicating different kinds of vertebrate and invertebrate traces, as identified in the legend.

Low tide could leave a shallow isolated pond where fish are forced to swim very near the bottom, allowing them to leave sinuous fin trails called *Undichna*. Insect larvae meander in the exposed mud at low tide as well. The cyclic nature of tidal influences is impressed on the sedimentary rock that was preserved at the UCM. As we noted in chapter 5, however, it is unlikely that a visitor to the Minkin Site during the coal age would have seen all the animals depicted in figure 9.8 on a given day.

Studies of the rock layers at the Minkin Site have provided a great deal of insight into the prevailing environmental conditions when the tracemakers were alive. A scientific approach has clearly been needed to provide a reliable picture of the site. But the richness of the Minkin Site showed that there was much more that could be learned if only amateur collectors and professionals had enough time to explore what was there in greater detail. Somehow, reclamation had to be delayed. We now turn to the story of how the BPS (and its later offshoot, the Alabama Paleontological Society) made this happen.

10

Discovery in New Mexico

At the time of the Union Chapel Mine trackway discoveries, BPS members were unaware that fossil trackways of Permian age (280 million years ago) had been found in the Robledo Mountains of New Mexico by another amateur paleontologist, Jerry MacDonald. The discovery, made in 1987, caught the attention even of staff at the Smithsonian Institution in Washington, DC. The story of how and why MacDonald made his discovery, and what he did about it afterward, is beautifully described in his 1994 book *Earth's First Steps: Tracking Life before the Dinosaurs*. His remarkable experience was almost a how-to book for the next stages in the development of the UCM project, including salvaging efforts, documentation of the finds, and preservation of the site. Eventually, the BPS would invite Jerry MacDonald to come to Alabama to see the BPS discoveries firsthand. Here we give an introduction to Jerry MacDonald's work. In the next chapter we examine the similarities between his experiences and those of Steve Minkin in his eventual role of bringing the UCM discoveries to the attention of professional scientists.

Jerry MacDonald had been strongly interested in geology and paleontology long before he made his big discovery. Like most fossil hunters, he enjoyed what fossils represented: extremely old rocks that preserved the shapes of organisms that were once alive and that, more likely than not, were now extinct. He also enjoyed the desolate areas where people often find fossils, areas that may once have been tropical paradises. In 1983, he was tantalized by a fossil footprint exhibit he viewed in the Earth Science Department at New Mexico State University (NMSU), where he was enrolling as an undergraduate student. He was used to thinking of fossil footprints as being exclusively dinosaur footprints that would be too large for anyone but museum staff to collect. But in the NMSU exhibit he saw smaller tracks. MacDonald realized that physically, anybody could collect fossil footprints.[1] He wanted to know where these came from, but the display included no identifying information. He eventually learned that tracks were occasionally found at a rock quarry in the nearby Robledo Mountains, what he calls an "unassuming range" just west of Las Cruces. Several years went by before his curiosity about fossil tracks rekindled. In the meantime he received a bachelor's degree in earth science from NMSU, in 1985.

In the fall of 1986, after thinking a great deal about the tracks and what he had learned about them so far, MacDonald began to seriously entertain the idea of searching the Robledos for the elusive layers of rock the tracks must have come from. Initially, the search was enervating and unproductive. It was hard to find tracks in the Robledos. He could search for hours under the hot sun and not find a single track. At the same time, he was concerned about the quarrying operation in the mountains. It was destroying a layer of extremely well-preserved Permian plant fossils. He was also intrigued by buildings in Las Cruces made from some of the rocks from that quarry. A restaurant in Las Cruces had a patio floor made of slabs with trackways, and a local house was advertised as having floors with fossil ferns.

MacDonald's interest in fossil tracks really started to get serious when the local natural history museum in Las Cruces held a "bring in your fossils" day in the spring of 1987, with the idea of having museum staff look at fossils collected by amateurs for donation. He was one of those given the task of identifying the donated fossils. One was a large slab covered with small tracks, donated by a man who found it in a "wash" (dry creek bed) in the Robledos. That is, the specimen was not found in situ, but in a jumble of rock that had been deposited by a flash flood. MacDonald wanted to find out exactly where these small tracks were coming from. To add to the depth of his interest, he serendipitously found trackways when he and his wife, Pearl, visited the house advertised as having fossil ferns in the floor. He discovered that the "ferns" were really pseudofossils called dendrites, which look deceptively like fern fossils,[2] but as he looked around the house further, he found real tracks in the stone floor in the living room. The rocks had come from the Robledos. MacDonald began to seek out information in the literature about fossil plants and tracks in the Robledos and about well-known fossil trackway sites in other locales.

The literature on fossil tracks in the Robledos proved largely unhelpful in pinpointing the source of the tracks people had found. MacDonald recounts in the introductory paper to "Early Permian Footprints and Facies," a 1995 publication on his finds, that of half a dozen articles he found on fossil tracks and plants in the Robledos, only two provided any trackway locality information. He ended up concluding that "no large tracksite was ever unearthed as a result of searching for fossil footprints in the Mesilla Valley" (Lucas and Heckert 1995). He also noted that there was a general belief among Robledos track collectors that the tracks were unidentifiable, mostly because the specimens that had been found so far had been exposed for a long time and were badly weathered. The source of the trackways was thought to be "flagstones" either eroded out of the mountains or exposed by the active quarrying operation. But finding tracks in a quarry was akin to finding them in a surface mine: you have to search tailings, or piles of broken-up rock. MacDonald concluded from all that he learned about track collecting in the Mesilla Valley that it was haphazard, indiscriminate, and largely unscientific.

The Coconino Sandstone is the host rock for one of the best-studied Paleozoic trackway sites in the world. Located in the Grand Canyon area, the early Permian, 280-million-year-old site yielded hundreds of well-preserved trackways when first investigated back in the 1920s. MacDonald was influenced by the methods of Charles Gilmore, the Smithsonian paleontologist who was assigned the duty of excavating the tracks and preparing a possible open-air exhibit. Gilmore excavated overburden to expose long, in situ trackways, which were the kinds of trackways that MacDonald dreamed of finding in the Robledos. (No in situ trackways have been excavated at the UCM.)

The bulk of the tracks Gilmore found were those of vertebrates, mostly reptiles, and he assigned them to many different ichnospecies. MacDonald notes that Gilmore was a "splitter." A splitter classifies tracks based on slight differences, implying that each was made by a different animal species. In contrast, a "lumper" attributes different-looking tracks to fewer animal species, considering the differences to be due more to variable preservation or behavior than to a large number of different species. We have already noted (chapter 7) that T. H. Aldrich was also a splitter. Early studies of trackways were prone to splitting because no bones were found of any of the trackmakers. This was as true of the Coconino trackmakers as of the UCM trackmakers. Lumpers tend to describe ichnofaunas comprised of fewer similar species. This is ecologically more consistent with the diversity of living faunas.

Regarding the trackways found around the same time in Alabama, near Carbon Hill at the No. 11 Mine of the Galloway Coal Company (chapter 7), MacDonald noted that these were unique to the geology of Alabama, and that they greatly excited the scientists who first saw and studied them. He had seen the paper by Aldrich and Jones (1930), and a particular remark caught his eye: the concluding comment that the tracks had "very little benefit to anatomists and morphologists in the United States," presumably meaning that the tracks were not bones, and without bones you couldn't say much about what the animal that made the tracks looked like. MacDonald felt that this opinion was widely held in the United States and that it stalled trackway research for decades. He was convinced that trackways could provide important clues about the behavior of early vertebrate animal life.

MacDonald, fascinated by the discoveries of Charles Gilmore and Aldrich and Jones, and tantalized by the "scraps" of New Mexico tracks he had seen and occasionally found, wanted to find what he anticipated would be the "mother lode" of tracks in the Robledo Mountains. After considerable persistence and painstaking searching, and armed with as much prior information as he could gather from the recollections of others and from the literature, he found what he was looking for on June 6, 1987: a heavy slab with a set of five seven-inch-long tracks of what must have been a huge animal. This discovery slab had to be split into two halves to reveal the footprints (fig. 10.1) (MacDonald 1992). The left half shows the actual surface impression of the animal's footprints, while the right half displays the counterimpression or natural cast of these surface tracks, made after sand covered the surface tracks probably not more than a day later (MacDonald 1992). MacDonald attributed the tracks to an animal known as a pelycosaur, a large reptile-like predator that likely laid its eggs on land and was an imposing part of the terrestrial fauna during the Permian.[3] MacDonald was clued to the discovery slab because he saw tracks on the side that was exposed when he pulled

FIGURE 10.1 *Top*, Jerry MacDonald's discovery slab; *bottom*, Jerry MacDonald hauling a massive slab to his vehicle a half mile from the excavation site.

it out of a hillside. It was only after he split the slab and found the tracks inside that he knew he had found what he'd been looking for.

The discovery slab was only part of what was likely to be a much longer trackway. But MacDonald was now in a position to follow that trackway as far as he could, because he was able to isolate the layer of rock the tracks came from. What happened after this is legendary: he became a one-man excavating machine. He was not able to park his vehicle any closer than a half mile from the site, so he had to carry massive slabs on his back (fig. 10.1, bottom).[4] After considerable work, MacDonald was able to follow the trackway farther and found more tracks from the same animal. The trackway "walked" into a cliff 90 m (300 ft) high, where it could be followed no more.

In situ trackway collecting is fundamentally different from collecting in scattered rock piles. Once MacDonald had found the source of the tracks, he enlisted the help of volunteers to clear out scattered, loose rocks from the hillside in order to prepare it for *systematic excavation*. The hunt was no longer random: now it was a matter of delicately removing layers of sedimentary rock and opening them up like the pages of a book. It is important to note that the discovery was not based on naturally exposed trackway layers (like some dinosaur trackways). It was essential to split slabs to see the tracks.

To MacDonald it was highly exciting to be the first person to see the footprints of such a long-dead animal. When he split his discovery slab, it was like opening a book that had not been read for 280 million years. He was fascinated by what the slab really meant: the track layer he was seeing was once a shoreline. In the millions of years since the animal walked, sediment deeply buried that trackway and then mostly eroded away, leaving only a 90 m (300 ft) "burial shroud." He says, "I couldn't conceive of a more thrilling experience than to uncover the footprints of an animal that no other human being had ever seen. On the spot where I was now standing, 280 million years ago, this giant reptile had slithered past" (MacDonald 1994, 62).

MacDonald also remarked how, after seeing what he had actually found, he knew these were not the kinds of things a person simply takes home and keeps in a private fossil collection.[5] He said he changed from a "curiosity seeker to a research scientist," especially after he realized he could follow the discovery tracks farther into the mountain. Such a significant find of fossil tracks had the potential to tell a great deal about terrestrial animal life during the Permian Period, even more so than the very limited number of skeletons from that time. To exploit this potential, it was essential that the specimens be carefully documented and housed in an *institutional setting* in order to preserve them for long-term study. Most of the specimens MacDonald found went to the New Mexico Museum of Natural History and Science in Albuquerque; some were donated to the Carnegie Museum and the Smithsonian.[6]

The result of MacDonald's perseverance in the Robledo Mountains has been twofold. First, his work has led to the largest and most scientifically valuable collection of Paleozoic Era fossil tracks and traces in the world. Indeed, the collection is so significant that it has been officially named "The Jerry MacDonald Paleozoic Trackways Collection" at the New Mexico Museum of Natural History and Science. Second, MacDonald's work in the Robledos did not end with the discovery of the source of the tracks. For nearly two decades, he persisted in getting the track site protected as a *national monument*. The Robledo Mountains are now part of the Prehistoric Trackways National Monument, a designation that protects the trace fossil assemblage there from indiscriminate collecting and will facilitate study over a much longer period than might otherwise have been possible. As noted by Martin Lockley, who wrote the preface to *Earth's First Steps*: "The New Mexico Museum of Natural History and Science now contains MacDonald's huge collection and several researchers have reoriented much of their research focus to Permian tracks. Such is MacDonald's legacy" (M. Lockley, pers. comm. 2011).

The final lessons from MacDonald's example are these: "Never take anything for granted," and "Be persistent." Now we turn to the catalyst that led to the protection of the UCM mother lode for long-term study.

11

The Minkin Idea

In the 1920s, the collection of fossil animal tracks in Walker County was very much in the domain of professional geologists and paleontologists. We are not aware that amateur paleontologists at the time amassed large collections of tracks that they donated to museums, probably because the tracks were found in underground mines that were not accessible to amateurs. Strip-mining was not common in the 1920s, but by the 1990s it was the method of choice for mining relatively shallow layers of coal in Walker County. Strip-mining is what allowed amateurs to engage in serious collection of coal-age fossils. The irony is that it also opened the door to renewed interest in Alabama trackways after professional interest had languished for decades.

Between 1929 and 1999, coal mining remained active in Walker County, with new mines regularly springing up in areas where a seam could be identified and followed, and old ones discontinued when the coal was exhausted, mining permits expired, or the mining company went bankrupt. In any mine where the interval between the Jagger and Blue Creek coal beds was excavated, tracks were bound to turn up (especially in the Carbon Hill area). Thus, we imagine that over the years after the initial discovery, tracks were seen in other mines where the lower coal beds of the Mary Lee coal zone were exposed. Some people undoubtedly did collect slabs of tracks, but these either went into private collections or were traded for other fossils, and they were likely never seen by trackway specialists. Thus, no new knowledge came out of those finds.

Jim Lacefield brought attention to two recent cases that help put into perspective what happened with the Union Chapel Mine. In early 2000 he spoke to the BPS about a mine that was once in operation near the very small town of Kansas, Alabama (chapter 7), just a few miles west of Carbon Hill. In the early 1990s, he and others from the University of North Alabama had visited this site and found it to be rich in vertebrate trackways, including well-preserved surface tracks. The site was so productive that he thought it was actually a better site than the Union Chapel Mine. Many specimens were collected, with some going to the University of North Alabama for a coal-age display and others into private collections. Lacefield tried to get local paleontologists and geologists at the Alabama Museum of Natural History and the University of Alabama interested in studying the material at the time, but none specialized in trace fossils and no follow-up work was done. He also told how a large mine near Cedrum, Alabama, a few miles east of Carbon Hill, yielded hundreds of high-quality, mostly invertebrate trackways, found when several faculty and students from the University of North Alabama visited the site in the mid-1990s. Nearly ten years later, after the Union Chapel discoveries had rejuvenated interest in the tracks and while outside professionals were visiting Alabama to see them, Lacefield tried to locate the Kansas and Cedrum Mine specimens that had been collected in those earlier years. Other than what he had collected himself, none of the material could be found (Lacefield and Relihan 2005).

Why did the Union Chapel Mine trackways not suffer the same fate of being lost to science as the material from these other mines? It probably boils down to the actions of one person. In 2000, Steven C. Minkin (1947–2004), a local licensed professional geologist and relatively new member of the BPS, recognized the significance of the Union Chapel fossils as rare and potentially scientifically important and decided to *mobilize members of the BPS with the dual goals of photographically documenting the Union Chapel trackways and bringing them to the attention of professional trace-fossil specialists working at nearby institutions*. Although the BPS had been formed in 1984 to support the work of local professional paleontologists in Alabama, the group had never taken an initiative like this on its own. Because of Steve's role in this initiative, and the great things that happened as a direct result of it, the Union Chapel Mine was named in his honor after his untimely death in 2004. Here we

they lack academic training they are generally not qualified or inclined to write research articles about them or evaluate their true significance. Although some amateurs can be effectively self-taught, it is still a good idea to communicate with real specialists about an important find. A second reason for contacting professionals is the sheer mystique of the trackways. Why are they preserved in Walker County, and what animals made them? Answering these questions requires experts who are knowledgeable in sedimentary geology, and who are familiar with the trace fossils found elsewhere and with the behavior of the animals that likely made the tracks. Finally, it is professionals who are best positioned to judge whether a new trace-fossil site has global significance.

Not long after visiting the UCM for the first time, Minkin contacted Professor Anthony J. "Tony" Martin, a trace-fossil and dinosaur specialist at Emory University in Georgia. This was the first contact between the discoverers of the trace fossils at the mine and professional paleontologists. Minkin telephoned and e-mailed Martin and convinced him that the mine could be very important. Upon learning about the site, Martin involved several students in UCM trackway research. One of these, Nick Pyenson, is now a curator at the Smithsonian. Researching trace fossils at the Minkin Site was an important part of Pyenson's transformation from student to professional paleontologist. This alone shows the value of what Minkin did.

Getting in touch with Tony Martin led Minkin to Dr. Andy Rindsberg, whose work on gas-bubble impressions was described in chapter 9. Rindsberg is a trace-fossil specialist (one of only two in Alabama) who at the time worked at the Geological Survey of Alabama (on the campus of the University of Alabama). Martin told Minkin to get in touch with Rindsberg, because studying Alabama fossils was part of Rindsberg's job. If the site was important enough, Rindsberg could investigate on behalf of the state. Ultimately, Rindsberg's contributions to the project were comprehensive. These included a collecting trip to the mine, helping organize a workshop on the fossils found, participating in interpreting given specimens, meeting with local congressional representative Aderholt about legislation to protect the mine, helping edit, coauthor, and author papers in the Union Chapel fossil monograph (chapter 14), and writing other publications about the trace fossils at the mine. He also publicized the site among professional paleontologists through one-on-one communications and by organizing a theme session at a meeting of the Southeastern Section of the Geological Society of America in Knoxville. Perhaps most importantly, he worked to educate collectors about trace fossils, about scientific collecting and documentation requirements, and about stewardship of the site. A recurrent theme was the need to ensure that important and controversial specimens were preserved in safe, accessible locations (like museums). This way, they would be available for study by others. He provided guidance throughout the entire rescue, preservation, and documentation process.

Steve Minkin was the only member of his family with a passion for fossil collecting. Yet all of Steve's family visited the site at one time or another, including his eighty-year-old mother, who enjoyed splitting rocks and looking for fossils. Minkin also took his brother Bruce to the mine more than once, and on one visit, Bruce found twenty tracks on the first rock he split open.

Minkin devoted a huge part of his life, from 2000 to 2003, to visiting the site at least every month and pushing himself to find and do everything he could to protect the site from reclamation. An inveterate enthusiast, Minkin tended to push himself too far. One time when Minkin and his wife, Missy, visited the mine, Minkin did not stay hydrated. On the way home he suffered a stroke. He was life-flighted to UAB Hospital in Birmingham. The effects of the stroke did not last long, and Minkin was soon back out at the mine. On a later visit, Minkin misstepped near the top of one of the tall rock piles at the site, lost his balance, dropped his rock hammer with a clang, and in a cloud of dust slid all the way to the bottom, coming out bruised and cut (fig. 11.1). That accident resulted in a second trip to the hospital for treatment of a slow brain bleed. Some wished that Steve's family would have helped protect him from himself. But Missy explained what anyone who knew Minkin well would have said. "Steve would have never wanted any restrictions put on his activity there at the mine. He just . . . loved being there and finding those fossils. So it meant just a great deal [to] him . . . , and that's why it means so much to the family that the site is named after him. It just really touches our heart" (M. Minkin, pers. comm. 2009).

FIGURE 11.1 Steve Minkin at the Union Chapel Mine, sometime between 2001 and 2003, smiling just moments after sliding down the large rock pile at right in the background.

"TRACK MEETS"

The greatest thing that Steve Minkin did was initiate a series of unusual meetings where Union Chapel collectors brought all their finds to a single location for direct inspection by professionals and for documentary photography. These meetings became known as "track meets" and turned out to be a highly effective way of "seeing the whole" of the site. People brought hundreds of specimens to these meetings and laid them out on tables for everyone to see. Never before in BPS history had the amateur collectors come together in such a systematic manner both to show others what they had found and at the same time to have their finds documented.

Minkin laid out the idea for these meetings in an e-mail to coauthor Ron Buta on June 22, 2000. He said that he and Ashley Allen were "working on a plan to have everyone who's collected tracks at the UCM bring their specimens to the Homewood Library." Homewood Library in Birmingham was where the BPS held its monthly meetings. Minkin envisioned setting up a camera in a corner and having tables available where people could lay out their specimens for labeling and then photography. In the e-mail, he asked Buta to be the photographer, based on Buta's previous experience in astronomical photography.

Buta jokingly applied the term "track meet" to the proposed meeting in a reply e-mail dated June 25, 2000. He suggested that a better venue for the "track meet" was the Alabama Museum of Natural History in Tuscaloosa. Minkin must have liked the term "track meet," because he used it five times in his reply e-mail of June 27. He said, "I see the main objectives for the 'track meet' as a) to develop a photographic record of as many trackways from the UCM as is possible, and b) an opportunity for the academic expertise to see a large collection of trackways at one time in one location."

According to his brother Bruce, Minkin recognized the importance of the site right away and wanted to make a serious effort to document the diversity, high quality, and sheer volume, especially of vertebrate trackways found at the site. After further e-mail correspondence between Minkin, Buta, and ALMNH collections manager Ed Hooks, the first fossil track meet was set for August 19, 2000, in the Alabama Museum of Natural History. Six other such meets would follow over the next twelve years (the most recent as of this writing was on March 2, 2013).

In an e-mail message dated February 3, 2002, Minkin summed up his motivations for the documentation effort:

> What inspired me to suggest that the BPS assemble all the tracks the BPS collected and invite the academic community to study: I taught Geology for the University of Idaho in Idaho Falls. What made geology come alive for the students were the field trips. For the Historical Geology classes, I collaborated with several paleontologists I knew from the Idaho Museum of Natural History in Pocatello, Idaho. These paleontologists would take my classes to the field for collecting in various areas. Whatever we collected (under museum personnel supervision), was made available to the museum. Several very important finds were made this way for Cretaceous dinosaur and Pleistocene mammal fossils. This was a fine example of the collaboration between the scientific community and

responsible collectors. I realized that the UCM track fossils were extremely rare and that the academic community should be aware of the BPS collection.

"Track Meets" in Action

The motivation behind the track meets was that Minkin was worried that scientifically important specimens were being rescued but would never be seen or used by scientists. Instead these specimens would disappear anonymously into private collections and never be studied. That is what happened to much of the Kansas and Cedrum Mine material collected by University of North Alabama students and staff during the early 1990s, as we recounted earlier in this chapter (Lacefield and Relihan 2005). That material was never studied systematically. When collectors pass on, or their interests shift, many specimens might simply be discarded. Also, people may not remember to document exactly where they found their tracks, which greatly diminishes a specimen's scientific value even if it is of exceptional quality. In 2000–2001, members of the BPS were visiting the UCM regularly, collecting numerous beautiful specimens of vertebrate trackways and the associated ichnofauna. Typically, collectors took home everything they found and either displayed it in their houses or stored it in their basements. The fate of the new material could have been identical to that of the Kansas and Cedrum material. Minkin was determined to prevent this.

The first track meet Minkin organized had two components: the laying out of the specimens, and a series of talks aimed at getting the collectors to look at the fossils from a scientific perspective. Collectors brought their tracks to a common location (the Alabama Museum of Natural History in Tuscaloosa) and initially laid them out on a set of large tables. With a few exceptions, each specimen was photographed with a metric scale and inventoried with a sequential UCM specimen number, the name of the collector, and a tentative identification of the fossil or fossils on the slab. Once so labeled, the specimens were taken to the second floor of the museum and photographed. Two photographers were on hand: Larry Herr photographed small slabs while Ron Buta photographed large slabs, both under artificial lighting. The whole procedure—the laying out of specimens, inspection by professionals, labeling, and photography—was done in assembly-line fashion in order to get through the large collection efficiently. Both amateurs and the local professionals present gave talks during the afternoon. By design, the first track meet had an air of professionalism. It was a model way of bringing amateurs and professionals together to work toward a common goal.

The first track meet went so well that additional meets took place as new specimens were collected, although none of these had the talk component (fig. 11.2). The track meets encouraged a feeling of responsibility among the fossil collectors and made it easy for research scientists to locate existing specimens that were not yet housed in museums. They also provided an inventory of photographs that were uploaded to the web in the form of the online Photographic Trackway Database, described later in this chapter.[1] Scientists anywhere can now see pictures of thousands of fossils from the site. This has enabled scientists to decide whether any of the specimens collected from the site relate to their research; to make a list of specimens they want to see, including where the specimens are kept; and even to make some preliminary interpretations, all without leaving home.

The track meets did more than just inventory what had been collected. One of their main purposes was to give everyone the opportunity to see each other's specimens—not just to study them, but to appreciate them. It was interesting for each collector to see the totality of UCM fossil collections, which they had never been able to do for other sites they visited. Everyone acquired a much better understanding of what the site really had to offer. But even more, Minkin organized the track meets to encourage owners of particularly good or unusual specimens to donate them to science museums in the state (M. Minkin, pers. comm. 2009). At the time, this meant the Alabama Museum of Natural History in Tuscaloosa or the Anniston Museum of Natural History in Anniston. It was only later that the McWane Science Center became involved in the project. Many collectors were quite happy to donate specimens, especially because of the involvement of professional ichnologists in actively researching the tracks.

A large fraction of the track meet photographs, especially from the first several meets, are of vertebrate trackways. These trackways are

exciting in part for the reason given in chapter 5: the time frame of the Minkin Site tetrapods coincides with the development and almost explosive radiation of the world's first reptiles. In addition, it is easy to see how vertebrate tracks were formed. The trackmakers are more closely related to us than are the makers of any of the invertebrate trace fossils at the site. Also, it is so clear from looking at the tracks that a creature was *going somewhere*. The tracks of vertebrate animals look like they were made by real feet with toes. They are natural action shots. The BPS members who visited the site "tuned" their eyes mostly for vertebrate trackways. An appropriate search image is essential to finding fossils (or seashells or anything else that might be lying on the ground).

Because the fossil collectors were looking for trackways, that's mostly what they found. From a scientific point of view, however, it is important to collect a *representative sample* of the fossils at a site like the UCM, including not only vertebrate trackways but also other kinds of trace fossils. Andy Rindsberg and coauthor David Kopaska-Merkel collected such a sample of the entire fauna in March 2000, about six weeks after the BPS first visited the site. Surprisingly, there is not much money in the state budget for GSA employees to collect fossils. The GSA is not only responsible for almost all research connected with the natural resources of the state; it is also a service organization. Citizens, businesses, and other governmental agencies often ask for routine help siting

FIGURE 11.2 A scene from Track Meet 3, which was held at the Anniston Museum of Natural History in May 2001. Specimens laid out on tables are being carefully labeled by Birmingham Paleontological Society members. BPS president Kathy Twieg is facing right in the foreground.

water wells or dealing with new sinkholes. Still, even at the beginning, it was obvious that the trace fossils from the Minkin Site were remarkable and possibly unique. And there was not much time to waste. It was thought that the mine would be reclaimed soon, probably within months, and it was important to preserve a suite of specimens for posterity. The specimens collected for the paleontological collection of the Geological Survey of Alabama on that quick expedition (fewer than one hundred) might have ended up the only ones readily available to scientists. It did not work out that way, thanks in large part to Steve Minkin.

One thing that helped with the effort to document and eventually protect the Minkin Site was visitors. In particular, well-known and respected ichnologists outside Alabama lent the weight of their reputations when BPS members tried to convince government officials and others of the importance of the site. One of these visitors was Hartmut Haubold, professor emeritus of paleontology at Martin Luther University in Halle-Wittenberg, Germany. Another was Spencer Lucas, curator of the New Mexico Museum of Natural History and Science. And most important of all, the BPS was privileged to have Jerry MacDonald come to Alabama and speak to the group.

The BPS initially found out about Jerry MacDonald by accident. Ron Buta went to South Africa from July 15 to August 15, 2000, to collaborate on some astronomical research with Prof. David L. Block, a colleague at the University of the Witwatersrand in Johannesburg. On the last day of his visit, he asked Prof. Block if there was somebody on campus he could talk to about paleobotany, and he ended up meeting with a couple of paleobotanists at the Bernard Price Institute for Paleontological Research, located on the campus of "Wits." He met the director of the institute and told him about the Union Chapel tracks. The director went to his bookshelf and pulled out an orange paperback book titled *Early Permian Footprints and Facies*, Bulletin 6 of the New Mexico Museum of Natural History and Science. The editors were Spencer G. Lucas and Andrew B. Heckert. When Buta returned to Alabama he ordered a copy. In an e-mail to BPS members on July 20, 2001, Buta told the BPS about the book, calling it "a very interesting compilation of papers concerning a 'megatracksite' in the Robledo Mountains of New Mexico" and pointing out that "unlike our experience, most of the tracks were collected by a single individual in a very organized excavation of a layered hillside over a period of about two years." Buta also noted that "the discoverer of the site, Jerry MacDonald, makes some very interesting comments that I think can help to put our work into perspective. Apparently, our group was a whole lot quicker in recognizing the significance of UCM than people were in New Mexico." He concluded, "I see some parallels between what we are planning to do and what they did [in New Mexico]."

Thus, from January 2000 to May 2001, the main effort of the BPS was devoted to salvaging as many trackways as possible before the seemingly inevitable reclamation of the mine, and to documenting through photography and inspection as many of the specimens as possible. For no other BPS field trip site had such a thing been done before.

The Online Photographic Trackway Database

The first three track meets were held in three different venues: the Alabama Museum of Natural History (Track Meet 1; August 2000), Oneonta High School (Track Meet 2; October 2000), and the Anniston Museum of Natural History (Track Meet 3; May 2001). Together these resulted in an assembly of more than one thousand color pictures of tracks from the UCM salvaged by twenty-four collectors. The bulk of these were taken by Ron Buta, whose policy was to photograph not only the full view of a trackway or other trace fossil, but also multiple closeups of different parts of a specimen. Also, the emphasis of the track meets switched from covering only the best-preserved (or best-looking) specimens to photographing all material, leaving the judgment as to significance to later research.

This unique database consisted mainly of photographic prints, of which three copies were made: one set for the Alabama Museum of Natural History, one for the Geological Survey of Alabama, and one for the BPS. The photographs were taken either indoors using artificial lighting, or (in the case of Track Meet 2) in full sunlight. Photographs after Track Meet 3 were taken with digital cameras.

The online Photographic Trackway Database was prepared by Buta and was a logical follow-up to Steve Minkin's documentation effort. Although initially it was not possible to post many of the track meet photographs online because of limited disk space, by early 2002 it was possible to post the entire trackway image database on the Internet for the whole world to see. As noted by Atkinson, Buta, and Kopaska-Merkel (2005), "the instant availability of all these images via the internet to distant scientists, government officials, and media representatives turned out to be a key advantage when the decision was made the next summer to push for permanent protection of the site."

Now we turn to the delicate issue of how the Minkin Site was granted protection by the state of Alabama for long-term study. The fact that this happened within six years of the discovery of the site, when the same kind of process took nearly two decades in New Mexico, is one of the most remarkable aspects of the story. How was this done?

12

Saving the Fossil Site

By the spring of 2001, the Union Chapel Mine had yielded so many excellent specimens of vertebrate and invertebrate trace fossils that three track meets were needed to document them all. The amount of high-quality material collected by multiple visitors was so great that it was clear the site could continue to yield for years. Yet, looming over the whole enterprise was *reclamation*. After the strip-mining of an area is completed, the mining company cannot simply abandon the land and leave it in a broken-up state. The federal Surface Mining Control and Reclamation Act of 1977 requires that the mined land be rehabilitated, meaning the rock piles should be smoothed out and the area regraded to eliminate any dangerous rises, like a highwall. The land should also be covered with soil, and vegetation should be restored.

Reclamation is a standard obligation of surface mining, and all mining companies in the United States are required to post bonds with their state surface-mining commissions in order to ensure that this obligation is met even if the mining company goes bankrupt before the restoration is completed. Reclamation is a good thing, because dormant surface mines leave a scar on the landscape and can also be hazardous for years after mining operations have ended. When a mine is fully reclaimed, the land can once again be used for a variety of beneficial purposes. In general, it is in everyone's best interest if the land is restored as closely as possible to what it was like before the mining started.[1]

In late 2000, mining at the UCM was finished, reclamation was scheduled, and by February 2001, the mine was already partially reclaimed. Although no one in the BPS disputed the need for mine reclamation, they hoped that an exemption might be allowed in the case of the UCM, the idea being that good laws like the Surface Mining Control and Reclamation Act of 1977 might be waived in a rare and exceptional circumstance. Steve Minkin believed that the UCM merited such a waiver. Concerned that such a rich fossil trackway site would be lost before its full potential to science could be realized, Minkin suggested to fellow BPS member Prescott Atkinson in early summer 2001 that a serious preservation effort be mounted. With Minkin's help, Atkinson took the lead on what was to be his most challenging experience as a collector of UCM fossils. The key components of the preservation campaign were the preparation of the online Photographic Trackway Database (based on the three track meets); support from the professional paleontological community, including citizen scientists like Jerry MacDonald; preparation of a major publication on the site (the Union Chapel Monograph; chapter 14); support from the mine owner and local and national government entities, including conservationists; and a flurry of newspaper and magazine articles highlighting the discoveries and bringing attention to the preservation campaign and what was at stake if it failed.

One advantage the BPS had at the start of the process was the support of the owner of New Acton Coal, Dolores Reid. She wanted to cooperate with fossil enthusiasts and scientists interested in the trace fossils. It was her inclination to allow the fossils to be collected, studied, and displayed for the benefit of everyone. Although not trained as a scientist, she understood the importance of the fossils found in her mine to both the scientific community and the public at large. To paraphrase Walter B. Jones (chapter 7): Dolores Reid and New Acton Coal had "fostered a notable contribution to science. Without this cooperation, the discovery could never have received the attention warranted by its importance."

The convenient location of the Union Chapel Mine some 55 km (35 mi) west of Birmingham made it easy for collectors to visit the site frequently (accounting for more than 2,400 specimens by mid-2005). The emergency salvaging phase took place during the first eighteen months after mining operations ended, before specimens could degrade significantly from exposure and before full reclamation activity was initiated.

At first, the BPS made no headway in rescuing the site because, even though the mine's owner supported the effort, mine reclamation is a federally mandated obligation, as we have noted. There are good reasons for this, connected with public safety and protection of the environment, especially water resources. Still, the site deserved an exception. What happened next was, as Ashley Allen recalls, "kind of a roller coaster ride. . . . You get a phone call one day saying, hey it looks positive, we got some folks involved who can make some things happen. You get a call two days later saying, oh, they're going to start reclamation in a couple of days and the site's going to be gone forever, and a couple of days later you get another call, and oh, hey guess what? We got a stay of execution on the site" (A. Allen, pers. comm. 2010). Prescott Atkinson noted that "we [the preservation effort] were dying on the vine until the press got hold of it. The press loves conflict. If you can present them with impending doom . . . famous fossil mine, largest Carboniferous track site about to be obliterated . . . perfect thing for an article. When Andy [Rindsberg] put [the word out] on Paleonet [an e-mail list for professional paleontologists] and the media got hold of it, then we got people's attention" (P. Atkinson, pers. comm. 2009).

JERRY MacDONALD AND THE PRESERVATION EFFORT

An essential part of the process of preservation was to submit a petition to the Alabama Surface Mining Commission in Jasper, asking it to consider the request. The petition was submitted in September 2001, and although the BPS made a compelling case, the prospects of success seemed bleak. On October 20, 2001, Atkinson e-mailed Buta that he thought the petition was likely to go nowhere, primarily because of the liability issue caused by the highwall. Everyone agreed that the highwall, a vertical cliff more than 20 m (65 ft) high, was dangerously unstable. Still, the site was too significant to allow this to kill the preservation effort. Outside support was needed, and Buta suggested to Atkinson that, as a group, they should contact none other than Jerry MacDonald. Buta wrote to MacDonald on November 4, 2001, and received a reply on December 10:

Dear Ron,
It was with great excitement that I read your letter and saw the

accompanying photos. I have always believed that the Alabama tracks in general were overlooked and understudied. I am so pleased that you and the BPS are so enthusiastic about them.

These tracks are of great importance, not only for understanding your region's past history, but in relationship to the Paleozoic trackway occurrences in the Western United States, particularly with regard to track occurrences in the Coconino sandstone in N. Arizona and in the Abo/Hueco member in south-central New Mexico. For the first time we may have the opportunity to follow the emergence and development of a number of important faunal groups that, in their own right, are little known or understood.

In my opinion, the fact that they are Pennsylvanian in age is actually better from a research standpoint than if they, also, were Early Permian. Their value is not diminished by the fact that they are found in spoil piles. This was somewhat the case before my work here in New Mexico, but it is no longer the case now.

I have over 80 sites spanning about 150 miles of shoreline, all of which preserve a wide range of information, from the depositional environment that preserved the tracks, to co-habitation, food pyramid studies, invertebrates, plants, and other sedimentological features. Yes, the fact that they can all be found and studied in situ for such a great distance is what has made the Robledo Track sites the so-called Rosetta stone for dating and analysis of Paleozoic tracks not only in North America but also in South Africa, and Europe, particularly Germany. But now we can correctly fill in gaps that may have proved too far-fetched or speculative to consider prior to the Robledo discoveries. And this is precisely why the track sites you have found are so important. They may indeed be the best from that time period anywhere, but as a cross-correlative discovery they are priceless when studied with the Robledo tracks. It is not necessary that they be found in situ (of course that would be better) but it is no longer a research handicap. . . .

I look forward to continuing correspondence.

Jerry MacDonald
Paleozoic Trackways Project

Later, the BPS invited MacDonald to see the Alabama trackways first-hand, and he accepted the offer to visit.

The trackways MacDonald found in New Mexico are Permian. They are a few tens of millions of years younger than the material found at the Minkin Site. For years, MacDonald studied these trackways almost in isolation and fought to have them preserved for science and posterity. As we described in chapter 10, he was famous for his tenacity in excavating large slabs, carrying them a half mile to his truck, and then reassembling them. His initial findings generated disbelief: some even thought he might have etched the tracks on the rocks himself! He donated large sets of tracks to museums, including the Smithsonian, which would later help fund the Paleozoic Trackways Project.

MacDonald was very interested in Alabama tracks and eager to have the perspective provided by other sites. He was initially aware of only the Carbon Hill discoveries that Aldrich and Jones had described in 1930. As he related in a more recent interview: "The Alabama material [was] crucial to my experiences and my conviction that my stuff was significant, you would have to say it was fundamental along with Gilmore's Coconino stuff [at the Smithsonian]. . . so to find out then, what was it, 2000 I think, that there's been this discovery and new interest in Alabama, was just really thrilling to me. I just thought it was, you know, history repeating itself" (J. MacDonald, pers. comm. 2009).

During his four-day visit to Alabama in April 2002, MacDonald met with BPS members to discuss progress on the Union Chapel Monograph (chapter 14), which had begun only the year before. MacDonald also spoke at the monthly BPS meeting about his Permian track discoveries in New Mexico and examined dozens of trackways that members brought to the meeting. The BPS took MacDonald and his wife, Pearl, to the fossil site, which as we have noted is very different from trackway sites in the Robledos. In those mountains, the tracks are in place, meaning that long trackways (tens of meters or feet long) can be examined in depositional context. Despite the differences between the two sites, MacDonald said that "the substrate is different, and the age is different, but [the Minkin Site fossils] are still remarkably similar in many respects to the material that I'm finding. To me it's like an earlier chapter in a book that we found" (J. MacDonald, pers. comm. 2009).

MacDonald noted that every part of the fauna at the Minkin Site was important, including invertebrate trackways. The more completely the fauna was documented, the more diverse the information, and the better our understanding. He said, "The invertebrate tracks were incredibly helpful to [the] paleobotanists in helping to flesh out the type of biomass, the paleoenvironments that we had here [in New Mexico] and I think the same is going to be true in Alabama. And of course that's going to be tremendously important. . . . What I would like to see developed in Alabama is a really good picture of what the environment was like, and the habitats, and the different animals, and the trophic structure [who ate whom], what was going on there" (J. MacDonald, pers. comm. 2009).

MacDonald's experience in New Mexico was clearly relevant to the preservation effort. Here was someone who had spent a significant fraction of his life collecting, documenting, and making available to professional scientists the fossil traces of animals that had lived in his state 280 million years ago. He had had success in getting local professionals to support his efforts to bring the Permian tracks into the research domain. Most of all, he appreciated the significance of the Alabama material and supported the effort to protect the mine.

A Scientific Gathering

Given the level of professional interest in vertebrate trackways found at the Minkin Site, it was inevitable that a scientific meeting to showcase research about the track discoveries and their global context would be held. The Permo-Carboniferous Workshop (figs. 12.1 and 12.2) was organized in 2003 to allow researchers from different institutions to interact and to make oral presentations about what they knew. The Alabama Museum of Natural History hosted the workshop over three days: two days of lectures (including an opportunity to examine many of the best specimens) and a day-long on-site field trip. Twelve technical talks and two posters were presented. The program book contained extended illustrated abstracts contributed by more than thirty professionals and students. The whole event cost about $1,000, most of which went into providing attendees with the workshop program, copies of the Aldrich and Jones booklet, and Jim Lacefield's *Lost Worlds in Alabama Rocks*.

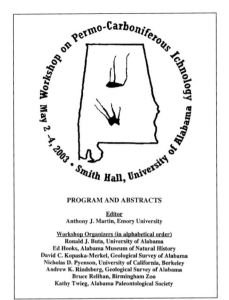

FIGURE 12.1 Program cover for the Workshop on Permo-Carboniferous Ichnology, held at the Alabama Museum of Natural History in May 2003.

FIGURE 12.2 Lauren Tucker from the University of Birmingham, United Kingdom, speaking at the Workshop on Permo-Carboniferous Ichnology.

The Permo-Carboniferous Workshop helped professional paleontologists from out of state learn how important the site really was. Scientists who participated, like Spencer Lucas and Adrian Hunt from the New Mexico Museum of Natural History and Science, Lauren Tucker from the University of Birmingham in the United Kingdom, and Molly F. Miller from Vanderbilt University, were impressed with what they saw. Some of these well-known paleoichnologists added their voices to the cause of protecting the site. Lucas and Hunt wrote a paper that was published in the Union Chapel Monograph (Hunt, Lucas, and Pyenson 2005), and Lucas also wrote a letter supporting the push to get government protection for the site.

The press release that accompanied the workshop shows how the organizers viewed the prospects for site preservation.

> The Union Chapel mine is in the process of being reclaimed. The piles of broken rock are beautiful to the eye of the collector but ugly to the casual visitor, and the crumbling 50-foot highwall is a standing hazard. The New Acton Coal Mining Company, which owns the mine and has been very generous to collectors, is now planning to collapse the highwall with explosives and cover the site with vegetation. The Alabama Surface Mining Commission has allowed a delay of reclamation but has no authority to halt the process, which is mandated under federal law. It will take an Act of Congress to save the site as a state park. . . . It seems ironic that the site that drew researchers here for an international conference will soon be covered with vegetation. . . . In New Mexico, Texas, Utah, and Connecticut, track sites have been saved as natural wonders for the public to see and enjoy and for researchers to study ancient life. . . . If nothing is done, Alabama will soon lose one of the world's best fossil sites, less than three years after it was discovered.

The Permo-Carboniferous Workshop did not result in a publication, but it brought professionals together to talk about their research on the trace fossils, and it helped the BPS realize that an independent volume about the site was both practical and desirable. Several workshop presentations were fleshed out and published two years later in the Union Chapel Monograph.

One way to protect the site for long-term study would have been for the US Department of the Interior to take ownership of the land, and perhaps make it a national monument. It appeared this could be accomplished only by an act of Congress. The Minkin Site was in what was then Representative Robert Aderholt's district, so any bill that would protect the site would have to come from his office. However, to have any hope of success with this venture, the BPS needed letters of support from outside scientists who could properly evaluate the significance of the fossil site. The question was, who to ask about this?

"The Most Important Discovery"

During his April 2002 visit, Jerry MacDonald gave Ron Buta copies of several major published works by Prof. Hartmut Haubold, a German vertebrate ichnologist at the Institute of Geological Sciences and Geiseltalmuseum, Martin Luther University (MLU), Halle-Wittenberg, Germany. One of these publications was a massive tome called the *Encyclopedia of Paleoherpetology*, published in 1971 (Haubold 1971; see also Haubold 1998; Haubold and Lucas 2001). In this volume, Prof. Haubold had collected all the information on vertebrate foot morphology and classification then available, including the Aldrich and Jones types found in Carbon Hill, Alabama. Prof. Haubold was clearly a world expert on fossil trackways, especially those from the Paleozoic, and the group needed his support.

Prof. Haubold was born in 1941 in Halle. His web page states that in 1959–60, he was an "excavation worker in the Hostage Valley brown coal." From 1960 to 1965, he studied geology and paleontology at MLU-Halle, receiving his diploma in 1965 with a thesis entitled "The Ichnological Fauna of the Red Sandstone of Southern Thuringen." In the European system, he did his second degree ("promotion") on the "Ichnology of Tetrapods in the Germanic Red Sandstones." He received his final degree ("habilitation," equivalent to a PhD) in 1978 with a thesis on "Biostratigraphy and Fossil Facies in the Thuringen Forest." He was appointed a university teacher and lecturer in paleontology at MLU-Halle in 1990. In 1992 he was appointed professor of paleontology. He retired from this position in 2006. On May 21, 2002, Ron Buta wrote the following letter to Prof. Haubold:

> Dear Dr. Haubold,
>
> Spencer Lucas may have informed you about a major fossil trackway find that has been documented recently in Alabama. I prepared a website which includes photographs of most of the specimens found and which highlights a monograph that is being prepared by

the Birmingham Paleontological Society, a group of local amateur collectors, in conjunction with several professional paleontologists, including ichnologists Anthony J. Martin (Emory University) and Andrew K. Rindsberg (Geological Survey of Alabama). If you haven't yet seen the website, the URL is http://bama.ua.edu/~rbuta/ monograph/.

The track site, known as the Union Chapel mine and located about 30 miles west of Birmingham, Alabama, is dated at 310 million years old (Early Pennsylvanian). You may already be familiar with the 1930 publication of Aldrich and Jones (Alabama Museum of Natural History, Museum Paper 9), who described vertebrate trackways from a nearby mine in Carbon Hill, Alabama. Incidentally, this work is still in print and we would be happy to send you a copy if you lack an original.

Dr. Lucas told me that you were interested in visiting Alabama next year during one of your regular trips to the US. We would be delighted to host you here to see as many trackways as you like and also to examine the site, an inactive surface coal mine in Walker County. About 4 weeks ago, Jerry and Pearl MacDonald visited us from New Mexico and spent one morning at the mine. BPS members have been collecting trackways from the mine since January 2000, and although no trackways have been collected in place, the gray shale spoil has yielded more than 1300 slabs/specimens with trackways, many of high quality, with the database including roughly equal numbers of vertebrate and invertebrate specimens. A few even show skin scales, according to Tony Martin and Andy Rindsberg.

Dr. Prescott Atkinson, a local pediatrician and active BPS member, has spearheaded an effort to preserve the Union Chapel mine. He has arranged to meet with a staff member of Alabama Congressman Robert B. Aderholt, a member of the US House of Representatives. The meeting will take place this coming Friday at 11am in Jasper, Alabama, not far from the fossil site. We are contacting members of the professional ichnological community to try and get letters of support for this preservation effort. Dr. Lucas has already agreed to write one. Would you also be willing to write such a letter? The letter should state why you think this site is important to paleontology, how it compares with other sites you are aware of, and why preserving it would be beneficial to science. I realize you might not be familiar enough with our project to write such a letter, and we will fully understand if it is not practical. Nevertheless, if you think you could write such a letter, Atkinson would need it by late Thursday night (5/23) or early Friday morning (5/24) at the latest, so I apologize for the short notice.

Many thanks for your consideration. I would be happy to communicate more information to you about the site and our documentation and preservation efforts.

Best Regards,

Ron Buta
Professor of Astronomy
University of Alabama

On May 22, 2002, Buta received the following reply (the letter to Representative Aderholt) from Prof. Haubold:

Dear Representative Aderholt:

I first learned about fossil footprints from Walker Co., Alabama at the end of the 1960s, when I wrote an encyclopedia on fossil tracks. A 1930 publication described the trackways found in an underground mine near Carbon Hill. At this time I inquired about the whereabouts of the specimens, and received an answer from the University of Alabama that the specimens described in 1930 are presumably lost. Last year I heard of the new discoveries of track fossils being made at the Union Chapel mine from N. D. Pyenson, an Emory University student who has been researching the trackways with Dr. Anthony J. Martin. On his request, I evaluated some of the discoveries to decide on their importance. I was in principal agreement with him that the new discoveries might be of world-class significance.

Now, during the last few weeks I was able to view the complete set of photographs of the new collection presented on the internet by The Birmingham Paleontological Society. *My assessment: by quantity, by quality, and by geological age, it is the most important discovery of Carboniferous tracks hitherto known* [our emphasis]. It will be the key for understanding the early evolution of tetrapod locomotion, and will solve globally the enigmatic understanding of tetrapod tracks of Pennsylvanian age. Besides the occurrences all over Europe I had the opportunity to study several Permo-Carboniferous track occurrences in the US since 1994. The Alabama site is of global paleontological interest. There is no comparable site in the world. And it would be the first extended site of Carboniferous age in the world that is protected.

Yours Sincerely,

H. Haubold

With this letter, we had the opinion of a distinguished vertebrate ichnologist who had written major works on fossil footprints. Even if the mine had not ultimately been preserved, this letter is a source of pride to all involved. In addition to Prof. Haubold's letter, the BPS also received strong letters of support from Jerry MacDonald, Spencer Lucas, Hans-Dieter Sues (former president of the Society of Vertebrate Paleontology), and José A. Gámez Vintaned (curator of the Paleontological Museum, University of Zaragoza, Spain).

Meeting Representative Robert Aderholt

The meeting with Representative Aderholt's staff members, Bill Harris and Paul Housel, took place as scheduled and was arranged by Prescott Atkinson. Atkinson convinced them that the representative should visit the site and listen to a presentation about why it should be protected. The visit took place on July 3, 2002, a beautiful sunny day, ideal for seeing and displaying tracks. Atkinson relates: "For the Congressman's visit to the site, we set up a tent and a table with a bunch of labeled tracks to show him the sort of things [we were finding]" (P. Atkinson, pers. comm. 2009). Figure 12.3 shows the representative at the site speaking with Andy Rindsberg.

Representative Aderholt sponsored a bill in the House to protect the site, but the bill never reached the floor. This attempt to protect the site by having the federal government take ownership failed. The failure of the Minkin Site to be designated a national monument is probably not, as some have suggested, because the Interior Department did not want to open the door to a flood of requests that other coal mines be so designated in order to save the cost of reclamation. Instead, we believe the bill died in Congress for reasons not directly related to the importance of the site, and the proposal to designate the Minkin Site this way never made it to the Department of the Interior.

Around this time, the BPS split into two groups. All the people heavily involved in the UCM project formed a new group called the Alabama Paleontological Society (APS). People working to secure protection for the site, primarily Prescott Atkinson, Steve Minkin, and Ron Buta, with the support of Bruce Relihan, then president of the APS, continued to research alternative preservation possibilities.[2]

FIGURE 12.3 Dr. Andy Rindsberg (*right*) speaking with Representative Robert Aderholt (*left*) about the Union Chapel Mine discoveries, July 3, 2002.

MEDIA REACTION

The media reaction to the discovery of the site, the rescue of specimens, and the attempts to preserve the site was a crucial part of the preservation effort. Media interest in the Union Chapel discoveries started with an August 15, 2000, press release about Track Meet 1, which was to be held the following Saturday (August 19) at the Alabama Museum of Natural History. The event was billed as a meeting "to examine footprints of some of the earliest vertebrates to walk on land, some 300 million years ago." Stories appeared the day after the event in the *Birmingham News*, the *Tuscaloosa News*, and the *Crimson White* (the University of Alabama student newspaper). But these general-interest stories were soon replaced by articles highlighting the rich value of the site and its impending loss to reclamation. Eventually, twenty-five articles about the site would be published in various papers, including important scientific magazines such as *Geotimes* and *Science*, and virtually all were sympathetic to the cause of site preservation.

One of these articles was written by Thomas Spencer and published in the June 5, 2003, edition of the *Birmingham News*. At that time, the APS had about two months to try to save the site. Prescott Atkinson nicely summed up the view of the APS when he said, "We've got something of world-class stature in this state, and we are about to just destroy it for no reason at all." The fossil discoveries had the potential to be a source of pride to Alabamians, a world-class fossil site in their own backyard with great long-term educational value, but reclamation still seemed inevitable even though the mining company was willing to donate the land. New Acton general manager Billy Orick was quoted as saying: "We're agreeing that we will donate the property if they can come up with a vehicle." In a later statement, on September 9, 2003, in the *Atlanta Journal-Constitution*, Orick said: "Personally, I'd like to see the site preserved. But an order is an order. It's going to take someone pretty high up to stop it [reclamation]."

Ed Howell wrote a particularly important two-part article for the June 18 and 19, 2003, editions of the *Daily Mountain Eagle*, the local newspaper of Jasper, Alabama. Part 1 was published the day before the APS was to meet with the Surface Mining Commission, and part 2 was published the day of the meeting. These articles laid out the issues involved and what was at stake. The first article quoted Prof. Hartmut Haubold, who argued that there was no comparable site like the UCM in the world, and that it would be the first Carboniferous-age site ever specially protected. Also in the article, coauthor David Kopaska-Merkel made the case that the site was the "oldest high-quality vertebrate track-way site known anywhere in the world." Spencer Lucas characterized the site as a "Rosetta Stone by which all similar sites may be interpreted." The article noted that both APS members and representatives of New Acton Coal, who supported the preservation effort, met with the commission at the same time. The irony is that the mining commission was initially taken aback by a group of people who were not involved with mining requesting that a mine *not be reclaimed*. In general, people wanted the land restored.

The Howell article goes on to discuss the bill submitted by Representative Robert Aderholt, Fourth District of Alabama, and the potential for the UCM to draw in tourism close to Corridor X, a planned road construction project (now I-22). As we noted previously, it appeared that an act of Congress was needed to circumvent the Surface Mining Control and Reclamation Act. New Acton permits manager Billy Orick was quoted as saying that if the company were released from the federal regulations and its permits, then it would donate the property. He also noted that the "regulators have bent over backwards to allow them [society members] to collect items." And indeed, collectors were given a great deal of freedom to explore the rock piles at the mine. But Orick emphasized that without the act of Congress, reclamation would have to proceed.

The second Howell article brought attention to the need for a special entity to take over the site if it were exempted from reclamation. This was indeed a very important issue. The coal company was willing to donate the land, but neither the APS nor the federal government could take charge of it. The article pointed to the Alabama Department of Conservation and Natural Resources as an alternative possibility.

On June 19, 2003, Representative Aderholt introduced the Union Chapel Fossil Footprint Site Preservation Act to Congress. The specific goal of the legislation was to allow the New Acton Coal Mining Company to donate the Union Chapel Mine land to the state of Alabama, and to exempt the company from the reclamation required by the Surface Mining Control and Reclamation Act of 1977. The company would also get back the bond it had previously posted to ensure that it would comply with the 1977 act. In a press release from Representative Aderholt's office, he is quoted as saying, "I am pleased to be able to help preserve this site . . . and look forward to seeing this legislation voted on and put into action. There's no doubt that we need time to carefully examine this site, so that we can really appreciate all that it has to offer us as we explore the past." In a follow-up story in the *Daily Mountain Eagle*, Ed Howell noted that "the Alabama Surface Mining Commission's lawyer said that . . . he would also draft a resolution for the commission in support of federal preservation status. Milton McCarthy of Birmingham, the commission's attorney, said that by law the land should have been reclaimed six months after the mine closed. He said the commission is still willing to be lenient for a while, although some type of action needs to take place."

The company received a letter in July 2003 directing that reclamation procedures begin immediately. The bulldozers were ready and all seemed lost, but the media took greater notice when Andy Rindsberg posted the impending loss on several well-traveled e-mail distribution lists. This attracted attention from *USA Today*, one of the most widely read newspapers in the country, which on August 20, 2003, carried an almost-full-page article on the site and the then-unfolding events relating to preservation versus reclamation. The article, written by Claire Bourne, outlined well the opposing forces involved. On the one hand, the Surface Mining Control and Reclamation Act of 1977 had no provisions for exempting a discontinued mine from reclamation. According to Dolores Reid, Union Chapel mining operations officially ended in 2000 because the coal had been largely exhausted in the area owned by New Acton Coal Company and it was not economically feasible to continue mining there. Under the law, reclamation had to begin soon after mining operations ended. The need for quick reclamation was compounded by a local resident who in August 2001 had purchased the rights to the land occupied by the mine to use for grazing his cows. Not only did he want New Acton Coal to honor the agreement to let him buy the land, but he was also worried about the potential hazard of the highwall left by the mining operation, a valid concern. The 1977 Surface Mining Act required that New Acton Coal prepare the land for sale, including removing the highwall hazard. Full reclamation also meant covering up the fossils and essentially losing them forever. Although by this time amateurs had salvaged well over a thousand specimens, this was only a small fraction of the trackways that could potentially be found.

As the period of uncertainty continued while the mining company appealed, more articles appeared. An excellent article by Mike Toner, on September 9, 2003, in the *Atlanta Journal-Constitution*, emphasized the special nature of the site and its exceptional richness in coal-age plants and trackways, including some that had never been previously documented, as well as the potential of the site to yield valuable material for decades to come. The article also quoted statements of well-known paleoichnologists.[3] These scientists emphasized the world-class nature of the site as a part of our national heritage that should be preserved.

On November 17, 2003, coauthor Buta was surprised to see an editorial about the endangered fossil site in the *Tuscaloosa News*. While the Alabama Surface Mining Commission in Jasper was deliberating the issue further, the editorial had this to say:

> Dull-eyed lead-handed bureaucracy is threatening one of the world's richest fossil discoveries at a strip mine in neighboring Walker County. The mine is on the brink of international acclaim for the wealth of fossils discovered there in the past few years. Unfortunately, the site is also on the brink of being bulldozed and dynamited because of the Alabama Surface Mining Commission's insistence that its owner comply with federal reclamation laws. In the balance are fabulous coal age fossils that led one researcher to proclaim the mine "the most important site in the world for its age." Fish, early invertebrates, insects, reptiles and even spiders left fossil traces and tracks in the mud of a freshwater estuary that covered the site some 310 million years ago. Nowhere else on Earth is there such a complete picture of life during that era, scientists believe. But federal laws dating to the 1970s require the now inactive strip mine to be returned as closely as possible to its original, unmined state. The mining commission ordered the owner this summer to bury the

site under a man-made hill. Backed by the Alabama Paleontological Society, the mining company is appealing. But state officials say it will take an act of Congress to win an exemption. In any other state, the Walker County strip mine would be considered a treasure, a magnet for study, research, and even tourism. The seeming inability of state officials to work with our members of Congress to save this unique resource suggests all the fossilization isn't confined to a strip mine: some of it is between the bureaucrats' ears.

Although this editorial raised the issues involved well, it was rather impolite to the mining commission people, whom the APS members had had the privilege of meeting and who respectfully listened to what they had to say the previous June. It was unfair to characterize them as fossilized "between the ears," so Buta sent the following letter to the editor (published on November 25, 2003):

Dear Editor:

As a member of the Alabama Paleontological Society, I was surprised to see an editorial in The Tuscaloosa News (11/17) detailing our effort to preserve a world-class coal age track fossil site in Walker County. Our desire to preserve this site has stemmed from its incredible yield of fossil footprints and other traces of pre-dinosaur animal life that once existed on an ancient tidal mud flat. The APS has done everything it can to document the fossils found and make them available to researchers worldwide (visit online at bama.ua.edu/~rbuta/monograph/). I wanted to make clear our view that the Alabama Surface Mining Commission has cooperated as much as the law allows with our efforts. This is the first time a group such as the APS has tried to prevent a surface coal mine from being reclaimed in order to preserve a fossil site, and the issue has to be dealt with carefully given the requirements of the Surface Mining Control and Reclamation Act of 1977. We very much hope the mine can be preserved for future study and research. If the mine is covered up, Alabama will lose an internationally known fossil site and a wonderful opportunity to broaden the knowledge of our state's natural history. We would also lose a potential tourist attraction and educational resource. But the APS members feel that the ASMC is doing its part to help cut the red tape.

THE FINAL PUSH

We do not know how much of an impact the published newspaper and magazine articles ultimately had on the final success of the preservation effort, but the company's appeal bought the APS enough time to explore a different approach. Although Representative Aderholt's bill did not pass, reclamation was delayed until a responsible entity could take over the land. Options included the APS itself, the Geological Survey of Alabama, and the federal Department of the Interior. As there were complications involved in giving responsibility to any of these entities, this left the Alabama Department of Conservation and Natural Resources (DCNR) to reach an agreement with the Surface Mining Commission allowing the State Lands Division to take over the site. The federal mine reclamation law includes an "escape clause" that allows state government to step in. This clause, never used previously in Alabama, waived the requirement for complete reclamation and allowed the state of Alabama to take ownership of the site.

The director of State Lands at the time was Jim Griggs, a former attorney for the Geological Survey of Alabama and the State Oil and Gas Board. He first learned of the UCM project from Nick Tew, the state geologist, who impressed on him the importance of the site and asked him whether the State Lands Division could take title to it. Griggs went to Jasper and met with the Alabama Surface Mining and Reclamation Division (SMR) to see whether a waiver was possible (J. Griggs, pers. comm. 2010). Eventually, the two agencies reached agreement. SMR determined that the mine did not need to be immediately reclaimed, and State Lands agreed to take over the site. Now it is available to collectors and researchers over the long term, and State Lands periodically turns over the surficial material to expose new specimens.

Reactions to the outcome included the following:

> "A producer with ABC News called us at school and I talked to him. . . . I had just found out the evening before . . . [that] we were successful, it's going to be preserved. . . . And we talked for a

little while, and I told him we would love for you to do a piece on that. . . . But other things were developing too, that sort of caught their attention." [And *ABC News* did not do a story on the mine.] (Ashley Allen, pers. comm. 2010)

The rescue effort was "a spectacular collaboration, . . . which certainly did result in the preservation of the site that in all likelihood wasn't going to be preserved. It wasn't even going to be known to the extent that it is now. If it wasn't for the BPS, [the rescue and preservation effort] would have never played out the way that it did. There's no way that a bunch of professional scientists, just acting alone, would have lifted this up to the level" that was achieved. (Jack Pashin pers. comm. 2010)

13

The Steven C. Minkin
Paleozoic Footprint Site

The formal transfer of land from the mining company to the Alabama State Lands Division took place on June 18, 2004. On March 12, 2005, the Union Chapel Mine was formally dedicated as the Steven C. Minkin Paleozoic Footprint Site, in honor of the late geologist's major role in and commitment to achieving the goal of preserving the site. The ceremony was held at the site from ten to eleven in the morning and was followed by a fossil hunt for all who attended. The ceremony was opened by Jim Griggs (fig. 13.1, upper left, on the right), who, on behalf of the State Lands Division, welcomed the attendees and unveiled the beautiful sign (fig. 13.1, upper right and bottom). The lead presentation for the APS was made by Prescott Atkinson (fig. 13.2), who had pushed hard for preservation after Steve Minkin's untimely death. Like Minkin, Atkinson had firmly believed it was possible to get the mine exempted from reclamation.

The speakers following Atkinson were Paul Housel from the office of Representative Aderholt; Berry H. "Nick" Tew, state geologist of Alabama; Randall C. Johnson, head of the Alabama Surface Mining Commission; Robert Reed Jr. of the New Acton Coal Company; and Ashley Allen and Ron Buta of the APS.

At the ceremony the APS also received a certificate for its effort in preserving the Union Chapel Mine. The certificate states: "The Alabama State Lands Division, Alabama Department of Conservation and Natural Resources, takes pride in awarding this Certificate of Appreciation to the Alabama Paleontological Society, in recognition and grateful appreciation of its commitment and dedication to the protection of Alabama's unique geological heritage through its untiring efforts in preserving the Steven C. Minkin Paleozoic Footprint Site at Union Chapel, Jasper, Alabama. In sincere gratitude, we recognize and commend these outstanding efforts to insure that Alabama's ancient natural history is preserved for future generations" (Buta, Rindsberg, and Kopaska-Merkel 2005a, 366). The document is signed by M. Barnett Lawley, commissioner, and Jim Griggs.

We are fortunate that some of the ceremony, as well as interviews with the speakers and others, was recorded for the award-winning *Discovering Alabama* series. Episode 56, titled "Tracks Across Time"

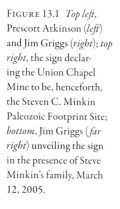
FIGURE 13.1 *Top left*, Prescott Atkinson (*left*) and Jim Griggs (*right*); *top right*, the sign declaring the Union Chapel Mine to be, henceforth, the Steven C. Minkin Paleozoic Footprint Site; *bottom*, Jim Griggs (*far right*) unveiling the sign in the presence of Steve Minkin's family, March 12, 2005.

(2005) gives an inspiring look at both the human and scientific sides of the story behind the Minkin Site. Expertly presented by series creator and host Doug Phillips, and written by Doug Phillips and Roger Reid, the episode provides insights from the people directly involved. Nick Tew noted in his speech that "a series of fortunate events led to the discovery of this site." Doug Phillips then commented on what these events were: "a student whose grandmother just happened to own a mine site, just happened to be in a science class, whose teacher just happened to be a member of the Alabama Paleontological Society, which just happened to recognize the significance of the trackways. And just happened to bring it to the attention of the proper authorities." But Phillips also observed that two other ingredients in the story were important: the depth of interest of the people involved, and their recognition of the value of the site.

Speakers Randy Johnson and Bob Reed were also interviewed. Dr. Johnson, the director of the Alabama Surface Mining Commission at the time, noted that for the past twenty-five years, his agency had had the task of getting mining companies to comply with the Surface Mining Control and Reclamation Act of 1977, "so you don't have these open pits and highwalls left." He further noted that the "issue that came up

FIGURE 13.2 Prescott Atkinson leads off the program at the dedication ceremony for the Steven C. Minkin Paleozoic Footprint Site.

with this paleontological site threw us a curveball." Coal mining companies like New Acton have to post bonds with the Mining Commission and must pay premiums on those bonds annually, to ensure that they will meet their reclamation obligations. Because of the fossil discoveries, New Acton delayed reclamation but still had to post the premiums at a time when the cost of reclamation was rising. Bob Reed, president of New Acton Coal in 2005, said the company had already started trying to reclaim the highwall when "these folks came in and discovered these fossils, so we stopped to see if there was actually some way to save this project." The APS was very fortunate that both of these gentlemen appreciated the significance of the site and were sympathetic to the cause.

Jim Griggs, also interviewed on the show, pointed out that the Minkin Site was only one of several sites the State Lands Division managed under the Forever Wild Land Acquisition Program, which allowed the purchase and protection of unique sites. "This is one of the earliest forms of wildlife represented at this site that we know, and it's as important that we protect the scientific data relating to that wildlife as it is that we protect today's unique places in Alabama," he said. He also noted that one of the missions of the State Lands Division was to educate young people to be future stewards of the land.

Other interviews in the episode provided scientific insight for the general public. Jack Pashin gave his take on what the Minkin Site was like during the Carboniferous Period: "What we think this was, was a coastal plain flanking the ancestral Appalachians. Here you have a mountain range forming in the southeast and you have rivers flowing out of that mountain range, and where the rivers approach the sea we have deltas and mudflats, and things like that. So we think that this was near the shore zone where the water was still very fresh and amphibians could live, but the area was in fact affected by marine processes enough that we could discern the tidal influence on sedimentation." Pashin also described how the rocks at the Minkin Site were dated based on their fossil content.

Coauthor David Kopaska-Merkel was interviewed at the site and noted that the sheer quantity of specimens allowed the potential of discovering rare traces: "When you have a lot of specimens, you find rare things," he said. Among such rare finds, he pointed out evidence at the

Minkin Site for fish swimming together and reptiles walking together, the earliest known examples of both schooling and herding (group or social behavior).

Prescott Atkinson summed up what led to the transition from a typical BPS field trip to the "track meets": "We realized after a few months had gone by that the volume of material we were finding was far beyond what we had the capacity as individuals to really handle. And we further realized that they [the trackways] were probably so abundant that we needed to bring them to the attention of professionals."

For Doug Phillips, the whole story was par for the course for Alabama and its remarkable natural heritage. "Alabama has a long history as a place of discovery," he said. The beauty of the "Tracks Across Time" episode is how it cements the five-year story of the nation's first-ever protected former coal mine site into the fabric of Alabama's educational history. This was a truly stupendous achievement for a group of amateurs, and the episode is a source of pride to all involved. "Stars in a sense are the fossils of heaven, taking us into the antiquity of the cosmos. But for all their shining glory, they have yet to reveal the most astounding manifestation of the Universe. They have yet to reveal the footprint of life. For that we have to look at a small but very significant plot of land in Alabama" (Doug Phillips, "Tracks Across Time," *Discovering Alabama*).

14

The Union Chapel Monograph

"No other track site has been documented in a comparable way yet" wrote Hartmut Haubold in a 2011 e-mail to the authors. The now-retired German professor of paleontology was answering a question about the publication titled *Pennsylvanian Footprints in the Black Warrior Basin of Alabama*, the monograph on the Minkin Paleozoic Footprint Site trace fossils that the APS had published in 2005. We were wondering what he thought of it now, almost six years later. Indeed, the Union Chapel Monograph stands as one of the signature achievements of the BPS/APS. Here we tell some of the story behind the preparation of this book, and how it has impacted paleontology in Alabama.

Almost from the beginning, we planned to prepare some kind of publication on the trace fossils being found at the Minkin Site by amateurs. Any publication the BPS might consider would have to be more than a guidebook and would likely have to include real research papers. When Ashley Allen spoke about the trackways at a BPS meeting, Ron Buta suggested producing a monograph about them. A monograph is typically a book written on a single specialized topic by a single author or by multiple authors. The Union Chapel Monograph is a monograph only in the sense that it is highly focused on a single fossil site. It is also an edited compendium of articles around a specialized theme.

Producing a book of this nature is usually a significant challenge. Someone has to step up to the plate as editor, collect thematic papers from other authors, and compile the book coherently. An editor may also author one or more of the papers. For the Union Chapel Monograph, sometimes referred to as the "Blue Book" because it was

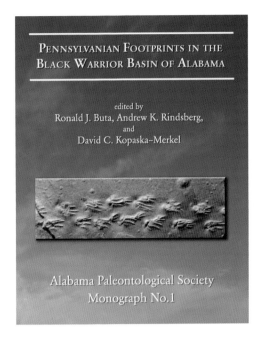

PENNSYLVANIAN FOOTPRINTS IN THE
BLACK WARRIOR BASIN OF ALABAMA

edited by
Ronald J. Buta, Andrew K. Rindsberg,
and
David C. Kopaska–Merkel

Alabama Paleontological Society
Monograph No.1

FIGURE 14.1 Cover of
the "Blue Book," or Union
Chapel Monograph.

published with a sky-blue cover (fig. 14.1), Ron Buta took the role of
lead editor and David Kopaska-Merkel and Andrew K. Rindsberg
agreed to be coeditors. As a professional astronomer, Buta had a suitable
background and had already coedited the proceedings of an astronomy
meeting held in Tuscaloosa ten years earlier.

Since Buta and others intended that the Minkin Site book be well
illustrated, they organized the "Great Track Layout" on July 27, 2001.
Photographs from Track Meets 1, 2, and 3 were laid out so local professional ichnologists could select the best and most scientifically useful
illustrations for publication. The gathering was also held to discuss the
research papers that should be included, the topics they would cover,
who the best authors might be, and how the papers should be arranged
in the table of contents. Most importantly, the event allowed collectors
and professionals to see the totality of the trace fossil database documented so far, constituting many of the nearly one thousand photographs taken up to that time.

At the time of the Great Track Layout, the Union Chapel Monograph was largely a pipe dream, an enormous challenge that involved

coordinating many people. The editors needed to (a) develop an outline, (b) contact contributors, (c) set guidelines, (d) manage manuscripts and reviewers (also known as "referees"), (e) apply publishing-format software throughout the book, and (f) seek funding for production and printing through the University of Alabama. Although the editors took the task on enthusiastically, the success of the venture was by no means guaranteed, dependent as it was on follow-through by contributors, potential funders, and so on.

Fortunately, the book was successfully compiled and published. It is divided into four major sections. The first includes three papers on the site's significance and discovery. In the first paper, Adrian Hunt and Spencer Lucas of the New Mexico Museum of Natural History and Science, and Nicholas Pyenson, then at the University of California, Berkeley, provide a global assessment of the significance of the Union Chapel *Lagerstätte* (a German word meaning "mother lode") and mention the site's "proven importance for public education." They also note that "the history of its preservation is unique in that a talented and diverse group of amateurs collaborated in the development and preservation of the site." In the second paper, Ashley Allen recounts his discovery of the site, while in the third paper, Ron Buta and Steve Minkin describe the salvaging and documentation effort (that is, what collectors were doing to find the fossils and how the "track meets" were being used to document the specimens).

The second section has eleven papers on the geology and paleontology of the site and provides the "meat" of the book, with discussions of environmental setting (Minkin), stratigraphic evidence (Pashin), vertebrate footprint taxonomy (Haubold and others), behavioral aspects of tetrapods and fish (Martin and Pyenson), a preliminary assessment of the invertebrate trackways (Lucas and Lerner), comparisons with other locales (Hunt and Lucas), larval insect traces (Rindsberg and Kopaska-Merkel; Uchman), arthropod body fossils (Atkinson), plant fossils (Dilcher, Lott, and Axsmith), and gas-escape structures (Rindsberg).

The third part has four papers on the impact of the Union Chapel Mine on both amateur and professional paleontology, including the importance of the amateur paleontologist (Hooks), ethics in fossil collecting (Rindsberg), how the mine was saved from oblivion (Atkinson,

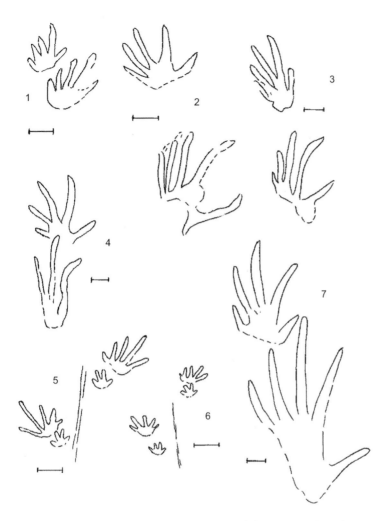

FIGURE 14.2 Manus and pes morphology of five ichnotaxa from the Union Chapel and Kansas Mines and the nature of their inferred makers. **1** *Notalacerta missouriensis* (amniote); **2–4** *Cincosaurus cobbi* (amniote); **5** *Matthewichnus caudifer* (amphibian); **6** *Nanopus reidiae* (amphibian); 7 *Attenosaurus subulensis* (anthracosaur or "coal lizard"). Each scale bar is 1 cm in length.

Buta, and Kopaska-Merkel), and the impact of the Union Chapel Mine project on Alabama paleontology in general (Lacefield and Relihan).

The final part includes three major atlases of vertebrate traces, invertebrate traces, and plant fossils. These include more examples of the different ichnotaxa and plant fossils than were covered in the main papers. Each atlas also has an accompanying table identifying the amateur collectors who found the specimens illustrated.

Figure 14.2 shows one of the main results of the Union Chapel Monograph: Hartmut Haubold's sketches of the five main vertebrate ichnotaxa from the Union Chapel Mine. *Nanopus reidiae*, a trace that had not been recognized before, was named after Dolores Reid because her extreme generosity made the whole UCM project, from Ashley Allen's initial scouting to preservation of the site, possible.

The monograph does not treat the interpretation of the invertebrate trackways in great depth. Nevertheless, the invertebrate traces at the Minkin Site are among the finest found anywhere, and much could be learned from them. For instance, many traces interpreted as being made by horseshoe crabs (and called *Kouphichnium*) have been reinterpreted by N. J. Minter (formerly at the University of Bristol) as being made by wingless, jumping insects called monurans. This is discussed further in chapter 16.

It took four years to produce the Union Chapel Monograph, which was modeled after the New Mexico publication on Permian tracks (Lucas and Heckert 1995). The final product, published in July 2005 with funding from the Geological Survey of Alabama Educational Committee, the McWane Science Center, and the Alabama Geological Society, represents a model of collaboration between amateur and professional paleontologists. Andy Rindsberg put it this way: "Production of a large, multi-authored book, with no grants funding the authors' work, within only five years of the discovery of the site is nothing short of a miracle." The book, when still in manuscript form, was used to good effect by Prescott Atkinson and others in the preservation effort.

The Union Chapel Monograph plays a special role in Alabama paleontology because it is the first major publication on Alabama's fossil footprints since Alabama Museum of Natural History Paper 9, Aldrich and Jones's pioneering 1930 paper that has been described in chapters 1 and 7. Eight decades ago, the discovery of fossil footprints near Carbon Hill caused a flurry of newspaper articles and attracted the interest even of paleontologists outside Alabama, but this single paper was the only significant attempt at researching the traces between 1930 and 2000. Because the Minkin Site is protected by the state of Alabama, and because amateurs are continuing to find new material more than a decade after the publication of the UCM monograph, interest in the trackways is still very high, and trackway research in Alabama should grow rather than languish as it did after the Aldrich and Jones study.

This interest will continue to grow because of Jerry MacDonald's efforts in New Mexico. The Jerry MacDonald Paleozoic Trackways Collection at the New Mexico Museum of Natural History and Science is one of the most significant paleontological collections of its type in the world and has brought Paleozoic ichnology to the forefront of paleontological research in the United States. Anyone studying that collection will be interested in the Alabama tracks and how those tracks and plants, the implied environmental conditions, and the behavior of the animals might relate to what is found in New Mexico, where the track-bearing rock layers are 30 million years younger than those in Walker County. The Union Chapel Monograph is a major contribution to the study of Alabama trace fossils, but it is not the last word on the subject. We expect that much more will come out of future studies of these fossils. In the meantime, amateurs will continue to play a major role in keeping the interest in Alabama trace fossils high.

15

Walkers and Swimmers

In this and the next chapter, we describe the trace fossils from the Minkin Paleozoic Footprint Site, beginning with the vertebrate traces, which are of two types: tetrapod trackways with rows of footprints, and fish trails, or sinuous traces made by the fins of fish swimming in shallow water.

Researchers do not agree on how many different tetrapods left their footprints at the Minkin Site. Martin and Pyenson (2005) argued in favor of only a single type of tetrapod trackway, *Cincosaurus cobbi*, which they interpreted as being made by an amphibian. Haubold et al. (2005a) identified five different trackway types (ichnotaxa), including both amphibian and amniote tracemakers. We use the classification of Haubold, who carried out the most extensive study of the tracks. The five different tetrapod ichnotaxa Haubold identified are: *Nanopus reidiae* and *Matthewichnus caudifer*, thought to have been made by amphibians; *Cincosaurus cobbi* and *Notalacerta missouriensis*, thought to have been made by amniotes, and *Attenosaurus subulensis*, thought to have been made by a large amphibian-like reptile (fig. 14.2).

According to Haubold, more than fifty primary ichnogenera (trace-fossil genera) of tetrapod footprints have been recognized worldwide. The classification of trace fossils, like that of body fossils, depends on recognizing distinctive characters that are unique to the type. For tetrapods, the main character for distinguishing amphibians from reptiles is the number of toe marks of the manus (forefoot) and pes (hindfoot). Although this sounds like a simple enough difference to recognize (see chapter 4 for an illustration of the difference), counting toes in a fossil trackway can be complicated by preservation issues, particularly the exposure of undertracks as opposed to primary surface tracks. If well preserved, the latter will show all digits and the classification will be obvious, but it is the former that are most likely to be preserved. As we showed in chapter 3, the farther an exposed undertrack is below the original track surface, the less information on actual foot morphology is preserved. For example, an animal may have five toes on its pes, but only three are commonly seen in undertracks. Other issues that determine track morphology are the animal's gait, consistency of the sediment, and what has happened to the trace since it formed. These all act to make tracks made by members of a single species look different from one another, such that no two footprints are identical, even when the feet that made them are.

The five ichnotaxa Haubold recognized were based on studies of well-preserved specimens that he was fairly certain showed manus and pes morphology recorded on or very close to the actual surface the animals were walking on. Because undertracks have a much higher preservation potential, only a few specimens out of more than a thousand collected were of the right kind to allow genuine recognition and classification of distinct genera and species of tracks. This shows the great value of the Minkin Site fossils. If surface tracks are very rare, then the probability of finding any at all is greatly enhanced when a given site is extremely rich in fossil trackway material.

While counting the number of manus and pes digits is useful for distinguishing amphibian traces from reptilian traces, other details can distinguish species within these groups. For example, foot morphologies may be slightly different for different amphibians, leading to different ichnospecies. A rich site can also be used for statistical studies. One can measure the length and width of a footprint, the pace and stride in a sequence of footprints (chapter 3), the orientation and shapes of digits, and the width of a trackway from right to left. With enough specimens, average sizes can be determined, showing that footprint sizes near the middle of the range are common while very small and very large footprints are rare. This is a typical result and is called a normal distribution. If only a few fossils are found, they may all look different, but if one has many specimens, the sample size will help show whether there are one, two, or more distinct groups (populations) based on size, shape, or other features.[1] Size differences, combined with observations like foot shape and differences in the number of digits on the forefoot, helped Haubold assign Alabama vertebrate tracks to five different ichnospecies. He did not perform statistical size analysis, although such a study could be undertaken in the future.

The lack of any body fossils of the animals that left the tracks at the Minkin Site makes it difficult to know what the trackmakers actually looked like. Body fossils of coal-age tetrapods from other parts of the world can provide some guidance (Clack 2012), while tracks provide information on gait and body length. The artist's renditions in this and the next chapter show our best estimates of what the trackmakers from the Minkin Site might have looked like. We used one of the best trackways to place the animal in a lifelike position. The two amniote ichnotaxa, *Cincosaurus cobbi* and *Notalacerta missouriensis*, were given roughly lizard-like heads, while the two amphibians, *Nanopus reidiae* and *Matthewichnus caudifer*, were given more rounded, amphibian-like heads. As shown by Clack (2012), coal-age amphibians and reptiles had a variety of skull morphologies, so picking the right one for the Alabama tetrapods is speculative. Only when actual fossil skeletons of Alabama tetrapods are found can we hope to know what they looked like. And even then we will have to determine whether such body fossils fit the trackways. This is part of the ongoing matching of feet to footprints that Martin Lockley has called the "Cinderella Syndrome."

AMNIOTES

Cincosaurus cobbi Aldrich and Jones 1930

Regarded as the "most famous [ichnotaxon] from the Pottsville Formation" (Haubold et al. 2005a), *C. cobbi* are narrow tetrapod trackways with pentadactyl (five-toed) imprints of manus and pes, both in reptilian-like arrangements (with curled or curved toes). These are the most interesting vertebrate trackways from the Minkin Site for two reasons: they are virtually uniquely Alabamian, and they were likely made by an early reptile. *C. cobbi* is found throughout the Black Warrior Basin coal fields, from Cordova to Eldridge. An excellent example is shown

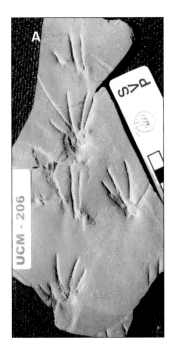

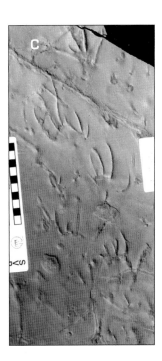

FIGURE 15.1 Examples of
Cincosaurus cobbi, tracks
of an amniote: (*A*) UCM
206; (*B*) UCM 254; (*C*)
UCM 451; (*D*) UCM 493.

in figure 8.4. Other examples are shown in figure 15.1, and evidence of group behavior is shown in figure 8.7.

The holotype, or "defining" example, of *C. cobbi* from Aldrich and Jones (1930) is shown in figure 7.3. This specimen, originally extracted from the ceiling of a mine tunnel, shows both manus and pes prints clearly but no obvious tail or body drag. The pes prints are larger than the manus prints and look like hands with thumbs, except that these apparent thumbs point outward from the trackway midline. There is also a thumb-like digit on the manus traces, and it too points outward from the trackway midline. If we were to get down on all fours and walk like an animal, our thumbs would point the other way: toward each other. Also, if the tracks of *C. cobbi* were made by an amphibian, we would see only four digits in the manus trace. Yet, more than one manus trace in figure 7.3 shows five clear digits. Thus, *C. cobbi* (which literally means "Cobb's five-toed lizard") tracks could not have been made by an amphibian. Instead, Haubold et al. (2005a) interpreted *C. cobbi* as having been made by an amniote, possibly a synapsid reptile.[2]

C. cobbi was only one of eight small vertebrate trace-fossil types named by Aldrich. The others were *C. fisheri*, *C. jaggerensis*, *C. jonesii*, *Quadropedia prima*, *Limnosaurus alabamaensis*, *Hydromeda fimbriata*, and *Trisaurus secundus*. Haubold et al. (2005a) suggested that these others were preservational variants of *C. cobbi*, because of differences in mud consistency or behavior, or because some were likely undertracks in which the appearance of the footprints had been modified. The other types are now regarded as *synonyms* of *C. cobbi*. Haubold had the benefit of the Minkin Site specimens to aid in this kind of reevaluation, which is common in paleontology.

Nearly all specimens of *C. cobbi* from the Minkin Site are undertracks, and the few primary trackways are poorly preserved, probably because the sediment was very soft. This is unfortunate, because *C. cobbi* is abundant and widespread (anywhere in the Black Warrior Basin where we find Pennsylvanian trackways, *C. cobbi* is abundant, and in western Walker County, it is typically the *most abundant*; chapter 19), and the name has been mistakenly applied to almost any trackway of roughly the same size. Interestingly, in undertracks, the full set of five toes on the manus can often still be detected, and these may terminate in a roughly circular footpad. In contrast, the pes is often reduced to two or three digits, which terminate in a roughly straight "heel." UCM 206 in figure 15.1*A* shows these characteristics.

In figure 15.2, an artist's rendition of a *C. cobbi* trackmaker shows it walking across a coal-age mudflat. (Note that the actual mud would have been a darker gray.) This is based on the footprints shown in figure 3.9 (UCM 263). In this case, the footprints come in pairs of barely over-lapping manus and pes traces, giving a large pace angulation and sug-gesting that the animal was not moving casually: it was "hustling" along at a fair pace. The tracks on this specimen are large (almost 4 cm [1.6 in] in length), and from the appearance of the pes, are probably under-tracks. Even so, we can use them to deduce the likely size of the animal, which was about 43 cm (17 in), or roughly the length of a Gila monster.

S.Blackshear

FIGURE 15.2 Artist's conception of the maker of *Cincosaurus cobbi* leav-ing tracks on the mudflat. UCM 263 (fig. 3.9) was used to scale the animal, which would have been about 43 cm (17 in) in length.

Notalacerta missouriensis Butts 1891

This is the second kind of trackway found at the Minkin Site that Haubold et al. (2005a) attributed to an amniote. It is much rarer than *C. cobbi*, being represented by only a few specimens from the Minkin Site and a few others from western Walker County. The two examples shown in figure 15.3 were collected from the Kansas Mine by J. Lacefield. The Minkin Site examples that have been identified are fragmentary and not well defined. *Notalacerta* is thought to have been made by an amniote because the gait is less sprawling than that typical of amphibians and because the manus has five toes (fig. 15.3, top panel, where a five-toed manus is seen in the middle, a quarter of the way in from the right).

FIGURE 15.3 Two examples of *Notalacerta missouriensis*, also made by an amniote. This ichnospecies shows a five-digit manus very similar to that of *Cincosaurus cobbi*, but it has a different pes morphology, with a long fourth digit. The two trackways shown are unnumbered specimens collected from the Kansas Mine in Walker County, Alabama, by Jim Lacefield.

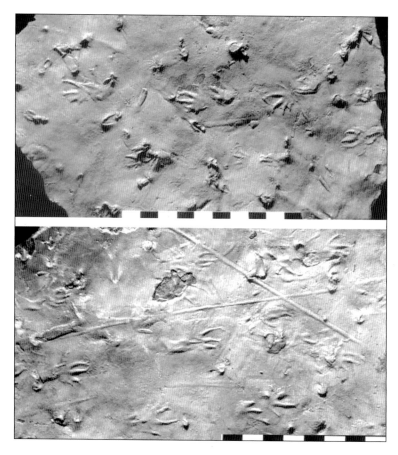

The digit lengths of the manus increase substantially from the first to the fourth (as shown in fig. 14.2), with the latter distinctly longer than the former. Trackway width and manus-digit proportions distinguish the species from *C. cobbi*. Chesnut et al. (1994) published important information about the characteristics of the ichnospecies.

Figure 15.4 shows an artist's rendition of *N. missouriensis*, using the trackway shown at the top of figure 15.3 as the model. These tracks indicate an animal only about half the size of the *C. cobbi* implied by specimen UCM 263 (fig. 3.9).

FIGURE 15.4 Artist's conception of the maker of *Notalacerta missouriensis* leaving tracks on the mudflat, based on the upper image in figure 15.3. The animal would have been about 20 cm (8 in) in length.

Amphibians

Although no proper statistical study has yet been made, the bulk of small trackways found at the Minkin Site appear to have been made by amphibians. In chapter 4, we noted that most coal-age amphibians were temnospondyls, animals having a characteristic opening in the palate (Clack 2012). Early Pennsylvanian temnospondyls were abundant and diverse. They included some of the largest amphibians that ever lived, but those that left tracks at the Minkin Site were no larger than salamanders. Haubold et al. (2005a) identified two ichnospecies in the Minkin Site trackway assemblage and considered both to be new to the Pottsville Formation in Alabama.

Matthewichnus caudifer Kohl and Bryan 1994

M. caudifer is the rarer of the two types. Individual tracks range from a few millimeters to slightly more than 1 cm across. The span from left to right feet ranged up to about 2 cm. The best example, UCM 285, is shown in figure 8.3. The tracks are distinctly different from amniote tracks: the manus is tetradactyl (four toed) and is much smaller than the pes (fig. 15.5, see also fig. 14.2). In the pes, the fourth digit

FIGURE 15.5 Example of *Matthewichnus caudifer*, attributed to a temnospondyl amphibian. Specimen UCM 469 (fig. 8.3). Note the small manus with only four digits, and the long fourth digit in the pes (arrow).

(counted clockwise from the left on the right side of the trackway, and the reverse on the left side) is much longer than the other digits. Figure 15.6 shows an artist's rendition of *M. caudifer* based on UCM 469 (fig. 15.5), a genuine surface trackway. The tracks imply an animal about 15 cm (6 in) long.

FIGURE 15.6 Artist's conception of the maker of *Matthewichnus caudifer* leaving tracks on the mudflat, using UCM 469 (fig 15.5) to scale the animal, whose length was about 15 cm (6 in).

Nanopus reidiae Haubold et al. 2005a

N. reidiae is named in honor of former New Acton Coal owner Dolores Reid, whose "generosity is arguably the single most important factor in the salvage of this large collection of tracks from the elements and from eventual destruction during reclamation." Like *Matthewichnus caudifer*, *N. reidiae* was likely made by a temnospondyl amphibian (fig. 15.7*A–C*). The holotype specimen, UCM 1142 (fig. 8.6), shows four-digit manus and five-digit pes traces. In undertracks, the manus may be weak or absent. Individual tracks are less than 1 cm across and trackways are roughly 1 to 3 cm wide. The manus imprints are only 60 percent of the size of the pes imprints. Along the trackway manus and pes imprints appear close together in sets with a changing pattern. The makers of *N. reidiae* were, on average, comparable to or slightly smaller than those of *M. caudifer*. This is shown by the artist's rendition in figure 15.8, in which an *N. reidiae* trackmaker approximately 15 cm (6 in) long is seen walking out of shallow water.

FIGURE 15.7 Examples of *Nanopus reidiae*, attributed to a temnospondyl amphibian. (*A*) UCM 357; note faint invertebrate trackway in upper left; (*B*) UCM 368; (*C*) UCM 312. The latter two trackways are typical of this ichnospecies.

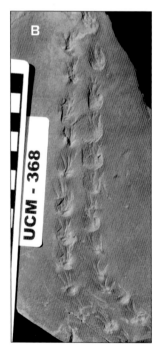

N. reidiae provides some of the most attractive and interesting small trackways from the Minkin Site, most often in the form of undertracks. Excellent examples are shown in figure 8.8, where two well-defined sets of *N. reidiae* trackways appear to cross, and in figure 8.9, where an *N. reidiae* trackmaker appears to have avoided an obstacle. Haubold et al. (2005b) show many more excellent examples.

FIGURE 15.8 Artist's conception of the maker of *Nanopus reidiae* emerging from shallow water and leaving tracks on the mudflat. The animal was placed into the tracks in figure 8.6 and would have been approximately 15 cm (6 in) long.

ANTHRACOSAURS

Attenosaurus subulensis Aldrich and Jones 1930

This species of vertebrate trackway, the largest found at the Minkin Site, is perhaps the most enigmatic. Haubold et al. (2005a) interpreted *A. subulensis* as the footprints of an anthracosaur (a protoreptile or amphibian-like reptile). The name anthracosaur means "coal-lizard." Many of the body fossils of these animals found elsewhere were associated with coal deposits of the Carboniferous Period. Anthracosaurs shared many characteristics with early reptiles, but their eggs, like those of modern amphibians, could presumably develop only in water. Amniote eggs (one of the defining characteristics of reptiles and their descendants) are enclosed by specialized membranes and, if they are laid outside the body, by protective shells, either leathery like those of turtles or hard like those of dinosaurs. The amniote egg, more than anything else, allowed true reptiles to colonize the land. One of the unique things about the Minkin Site is that amphibians, amphibian-like reptiles, and early reptiles all left their trackways in the same place. Several examples of *A. subulensis* tracks are shown in figure 15.9*A–D*. The largest known example, UCM 1621 ("Frogzilla"), is shown in figure 8.5*B*. *A. subulensis*

Figure 15.9*A–C* Examples of *Attenosaurus subulensis,* attributed to an anthracosaur ("coal lizard"). (*A*) UCM 1074; (*B*) UCM 24; (*C*) MSC 33968 (C. Wallace specimen).

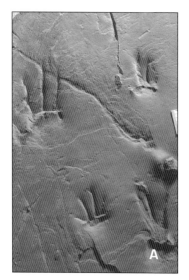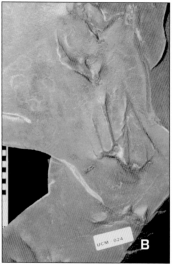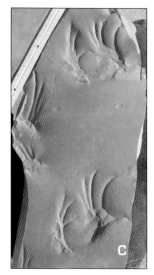

is not usually found with other vertebrate trace fossils (the cover is a rare exception), although tracks do occur with the ubiquitous larval insect trace called *Treptichnus apsorum* and other invertebrate traces like *Diplichnites* (chapter 16).

The typical *A. subulensis* trace involves one or two footprints rather than a full-fledged trackway, which would require at least five successive footprints to allow calculation of parameters such as pace and stride. Even so, a slab with only one or two tracks is of some use. For example, UCM 4085 and 123 (fig. 15.9*D*) each contain a single manus print that

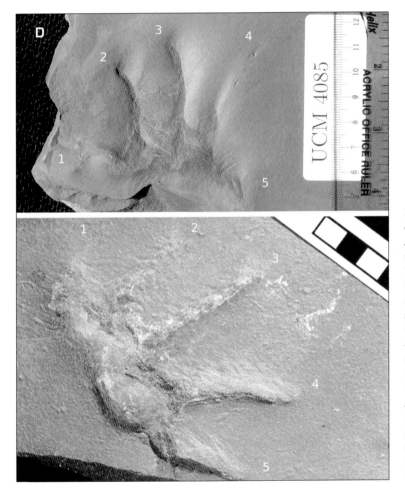

FIGURE 15.9*D* Examples of *Attenosaurus subulensis*, attributed to an anthracosaur ("coal lizard"). UCM 4085 (*top*) and 123 (*bottom*); The undertracks in UCM 1074 (*opposite*) show only four digits in the manus, while UCM 4085 and 123 both show a hint of the fifth digit (digits are numbered on the photos), showing that *A. subulensis* tracks were not made by a true amphibian. This ichnospecies includes the largest tracks found in Walker County (fig. 8.5*B*).

indicates that five toe digits were likely present. This means the maker of *A. subulensis* tracks was not just a big amphibian.

The original specimen from which *A. subulensis* was named has been recently located at the ALMNH in a damaged state, but numerous Minkin Site specimens are sufficient to document the characteristics of the ichnospecies. The only reservation that Haubold et al. (2005a) had about referring these specimens to *Attenosaurus* is the possibility that *A. subulensis* (rare large tracks) and *Cincosaurus cobbi* (common small tracks) were made by different-sized members of a single species. This is suggested by the occurrence of specimens of intermediate size and by other issues outlined below. Statistical analyses, which might answer this question, have not yet been attempted.

Minkin Site specimens referred to as *A. subulensis* by Haubold et al. (2005a) are larger (up to 25 cm [10 in] in pes length), exhibit a wider trackway pattern, and have different digit proportions than *Cincosaurus* (Haubold et al. 2005b). These differences may sound significant, but because the specimens from the Minkin Site are undertracks, they do not necessarily reflect the morphology of the feet that made them with great fidelity. It is difficult to compare them to specimens collected elsewhere. Problems with classifying and recognizing the makers of large Carboniferous tetrapod trackways were discussed by Haubold et al. (2005a), who concluded that the great size of the tracks assigned to *A. subulensis*, remarkably large for the Early Pennsylvanian, justifies naming and studying the taxon even though its fossil record is limited to individual prints and short trackways, mostly undertracks.

Hunt, Lucas, and Lockley (2004) carried out an independent analysis of the large tracks from the Minkin Site and divided them into three ichnotaxa: *Attenosaurus subulensis*, *Alabamasauripus aldrichi*, and *Dimetropus* isp., with most specimens being attributed to pelycosaurs, early sail-backed amniotes. This conclusion, that most large tracks at the Minkin Site were made by reptiles, which could lay their eggs on land, is different from that of Haubold, who identified the tracemaker as a creature that had to lay its eggs in water. This is an ongoing and unresolved scientific discussion. In the meantime we know two things: First, the oldest known pelycosaur body fossils are late Pennsylvanian. If tracks attributed by Haubold et al. (2005a) to *Attenosaurus* were made by pelycosaurs, then they would represent the oldest known pelycosaurs. And second, the *Cincosaurus* beds at the Minkin Site were deposited at the interface between land and water. Whether the large tetrapods living there were tied to the water by fragile jelly-clad (amphibian-like) eggs or could roam inland more freely, their role as top carnivores on the estuarine mudflats appears certain.

Figure 15.10 shows an artist's rendition of *A. subulensis* based on specimen UCM 1074 (fig. 15.9*A*). From this we estimate that a typical *A. subulensis* trackmaker was about 0.9 m (3 ft) long. The maker of the "Frogzilla" trace described in chapter 8 was likely bigger than this, perhaps up to 1.5 m (5 ft) in size, as big as a fully grown green iguana.

FIGURE 15.10 Artist's conception of the maker of *Attenosaurus subulensis* leaving tracks on the mudflat. The animal was placed into the tracks shown in figure 15.9*A* and would have been approximately 0.9 m (3 ft) long.

Fish Trails

While tetrapods were gaining a niche on land during the Pennsylvanian Period in Alabama, the oceans, seas, estuaries, and rivers were already teeming with fish. In fact, the major period prior to the Carboniferous, the Devonian, is sometimes called the Age of Fishes. Although no fish bones have been found at the Minkin Site, there is plenty of evidence they existed. Collectors have found many sinuous traces at the site. Such marks are made when fish, swimming in water not much deeper than their height, graze the bottom mud with one or more fins. Tony Martin suggests a reason fish would do this, creating the traces we find: "Fish don't normally like swimming along a sedimentary surface. It just impedes their progress. . . . In the Union Chapel deposit, a lot of those fish that were swimming along the bottom, it may have been because they were forced to. They might have gotten caught in pools at low tide, or otherwise shallow water" (A. J. Martin, pers. comm. 2010). Because fish bend their bodies in a wavelike (undulating) motion to propel themselves forward, their fins follow a similar pattern, which is why we see mostly sinuous trails. These trails are assigned to the ichnogenus *Undichna*, first described by Anderson (1976).

The way these trails relate to particular fins is beautifully illustrated by Seilacher (2007). A modern dogfish can leave as many as five sinuous traces, one pair from the pectoral fins, another pair from the pelvic fins, and a single trace from the caudal (tail) fin. If the pelvic and pectoral fins did not reach the bottom, or did not penetrate as deeply into the substrate as the tail fin, then only a single sinuous trace would be seen, in the form mainly of an undertrace. *Undichna* traces have been found in Early Pennsylvanian rocks around the world and also occur in deposits from later periods.

Out of 2,201 fossil slabs in the online database, 94 have *Undichna* traces (fig. 15.11*A–C*). Martin and Pyenson (2005) measured the wavelength (distance between successive wave crests) and amplitude (maximum sideways movement) of some of these trails and concluded that they were made by relatively small fish 10 to 15 cm (4 to 6 in) long. The shapes shown in figure 15.11*A–C* can be related to specific fins, swimming speed, and sideways tail speed. For example, when fish swim slowly, the grazing tail fin leaves a sinuous trace with a long wavelength,

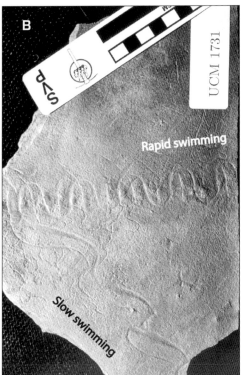

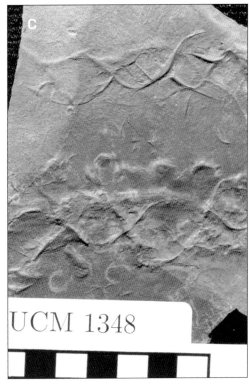

FIGURE 15.11 Examples of the ichnogenus *Undichna*, attributed to fish fins lightly touching sediment in very shallow water. (*A*) UCM 1728; (*B*) UCM 1731, two *Undichna* traces representing both fast and slow swimming; (*C*) UCM 1348. The specimen shown in *C* is considered strong evidence for schooling among coal-age fish. The out-of-phase double sine waves in *A* and *B* are likely the result of both the anal and caudal fins touching the sediment.

but if the fish speeds up and flaps its tail furiously, the trail changes to a shorter-wavelength (higher-frequency) trace (compare the two trails in fig. 15.11*B*). Most interesting are the short wavelength, high amplitude, and sometimes doubled traces shown in figure 15.11*A* and *B*. These kinds of traces suggested to Martin and Pyenson that the trailmakers were jawed fish that had both caudal and anal fins and used tail-based (as opposed to full-body or fin-based) propulsion to move around. In specimen UCM 1728 (fig. 15.11*A*), the higher-amplitude trace would have been made by the caudal fin, while the lower-amplitude, out-of-phase trace would have been made by the anal fin.

A small number (fewer than ten) Minkin Site specimens show multiple, roughly parallel *Undichna* traces that are not necessarily attributable to different fins of a single fish (fig. 15.11*C*). Instead, these appear to have been made by different fish moving together at the same time at a roughly constant distance, that is, in a *school*. Martin and Pyenson argue that this is the *earliest evidence of fish schooling in the geologic record*. Paleontologists always like to find the first, the oldest, or the largest specimens of anything. These are satisfying scientific landmarks and can also be important evidence of evolutionary milestones. Modern fish use schooling partly to confuse predators, and partly for social reasons. It appears now that this behavior may date back more than 300 million years.

Some *Undichna* specimens are more complex than those illustrated in figure 15.11*A–C*. Figure 15.12 shows a trace that appears to record a fish swimming around an obstacle in shallow water. When the fish approached a stick embedded in the mud it apparently veered sharply to its right and then less sharply to its left to avoid striking the stick. The high-amplitude asymmetrical fin trace indicates movement from right to left (Anderson 1976). The traces of at least three, and possibly five, different fins are represented in this specimen, each indicated by a separate color in figure 15.12. It is not clear whether this specimen of *Undichna* was made by a different species of fish than those responsible for most Minkin Site *Undichna* specimens. This specimen shows how valuable *Undichna* trace fossils can be for understanding the behavior of fish during the coal age. A more thorough analysis of *Undichna* is possible than has yet been attempted on the Minkin Site material.[3] Figure 15.13 shows an artist's rendition of a jawed fish leaving an *Undichna* trail.

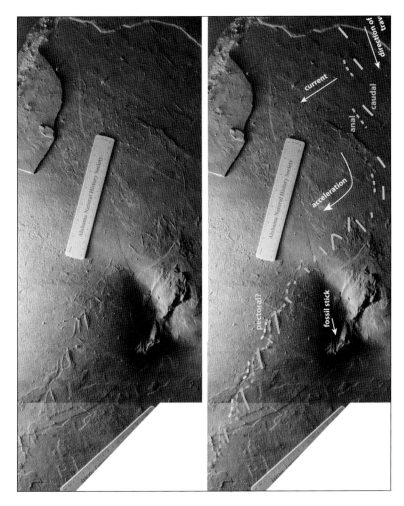

FIGURE 15.12 Complex and long *Undichna*. In this case, a small fish appears to be avoiding an obstacle in its path. *Left*, uninterpreted photograph; *right*, interpreted photograph, positive hyporelief.

What kinds of trace fossils co-occur with *Undichna* traces? Of the ninety-four slabs bearing *Undichna* in the online trackway database, twenty-three also include burrows made by fly larvae (dipterans). Tetrapod traces are generally not found on the same horizons as the fish trails. This is consistent with occurrences of *Undichna* in other deposits around the world (Soler-Gijón and Moratalla 2001; Turek 1989). The coexistence of larval burrows and *Undichna* traces on the same horizon of a slab does not mean that the larvae were burrowing just below the

FIGURE 15.13 Artist's conception of fish leaving *Undichna* trails at the Minkin Site.

sediment surface *while* fish swam by overhead. This is because in an estuary, the tide would have ebbed and flowed, with the fly larvae burrowing in the mud at low tide and the fish swimming just above at a higher tide, in a cycle that would have repeated once or twice daily. Thus it is not surprising to find examples in which larvae burrowed into surfaces that fish had already swum over, or vice versa. Because fly larvae preferred subaerial life, the dipteran traces on slabs with *Undichna* had to have been made at low tide.

Undichna is an important part of the vertebrate trace fossil assemblage at the Minkin Site. These specimens provide all the evidence we have about the aquatic vertebrates of the ancient estuary.

UNUSUAL VERTEBRATE TRACES

A database of trace fossils as large as that from the Minkin Site can yield very rare or unusual examples. It is important to look at some of these, as they can provide insights into some aspects of behavior or the environment that regular specimens cannot.

UCM 26 (fig. 15.14*A*) appears to record a small tetrapod emerging from a burrow and walking away. Haubold et al. (2005a) suggested that the tetrapod was a small temnospondyl amphibian, possibly the maker of *Matthewichnus caudifer*, and that the burrow (the large curved structure) was used for resting or estivation. Estivation in cold-blooded animals is like hibernation in mammals; amphibians estivate in winter in modern-day Alabama. Haubold also proposed an alternative interpretation: a hatching event. The footprints are smaller than those of other *M. caudifer* specimens, which could indeed mean that the trackmaker was a juvenile. However, this interpretation implies the absence of an amphibian larval stage. Haubold considered neither explanation fully satisfactory, and the specimen remains unexplained.

UCM 677 (fig. 8.9) shows a small tetrapod (*Nanopus reidiae*) walking around an obstacle consisting of several parallel linear striations. Martin and Pyenson (2005) interpreted these striations as being made by a partially buried horseshoe crab. This specimen is one of several from the Minkin Site that record interactions between living organisms and other creatures or inanimate objects.

UCM 680 (fig. 15.14*B*) shows a tetrapod surface trackway that was obviously made in very soft, wet mud. The animal dragged its body through the mud, leaving a mess! The median furrow (central section of the trackway) is broad and irregular, because bodies and tails tend to wallow in watery surface mud. This kind of body impression is not found in undertracks, because the marks made by torsos are lightly impressed. Tails drag harder, and feet push down hardest of all (though not equally between the heels and every toe). Surface trackways like UCM 680 are rare at the Minkin Site.

Some traces fit neither the sharp but diminished undertrack model nor the messy surface-track model. UCM 1034 (fig. 15.14*C*) shows an unusual diffuse trackway. These are tetrapod trackways in which the footprints have blurred boundaries but the central furrow lacks the sloppiness it would have if the mud were soupy (as in UCM 680). Tail drag marks are deeply impressed, which would not be the case if these were undertracks. Blurred track boundaries must therefore develop when tracks form in relatively coarse, dry sediment. For instance, tracks in dry

FIGURE 15.14 Unusual vertebrate trackways: (*A*) A small temnospondyl amphibian may have emerged alive from the mud in this specimen (UCM 26) and then walked away. It could have been intentionally "resting" in the mud (or hiding), or was buried during a storm but not too deeply. (*B*) A small amphibian sloshed in very wet mud, leaving a broad body impression and ill-defined footprints (UCM 680). Note the invertebrate trackway (*Diplichnites gouldi*) crossing the amphibian trace near the bottom of the image and extending to the upper right, and the invertebrate burrow (*Treptichnus apsorum*) crossing the upper part of the amphibian trackway. (*C*) Rare diffuse trackway (UCM 1034) with tail drag. Clearly a primary surface trackway, it may have been made in mud that was drier than usual.

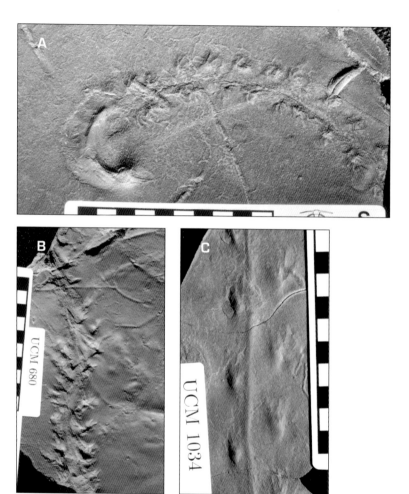

beach sand, like those of beach mice, invariably have diffuse boundaries. The diffuse tracks collected at the Minkin Site may have been made in dry sediment, but this would be very unusual in the swampy environment of the area.

This chapter has revealed the richness of vertebrate trace fossils that have been found at the Minkin Paleozoic Footprint Site. The tetrapod trackways are arguably the most important trace fossils from the site and so far have clearly proved of greatest interest to professionals and amateurs. Nevertheless, the invertebrate trace fossils have much to offer for further understanding of life during Alabama's coal age. Let us now turn to those.

16

Crawlers, Jumpers, and Burrowers

Some of the most beautiful and exotic trace fossils found at the Minkin Paleozoic Footprint Site were made not by vertebrate animals, but by invertebrates, such as insects or horseshoe crabs. In this chapter, we describe these fossils, which are of several types: *trackways*, made by animals walking on a substrate of wet mud and leaving trails of regularly spaced appendage (leg) marks; *jumping traces*, sequences of repetitive patterns made when an insect traveled by jumping; *resting traces*, where an animal left a single impression when it was "resting" in the substrate; and *burrows*, where an animal dug into the substrate, leaving tunnels that could be filled and preserved.

As noted in chapter 14, the invertebrate ichnofauna at the Minkin Site did not receive the same level of attention in the Union Chapel Monograph as did the vertebrate trace fossils. Had the site not included so many vertebrate trackways, collectors would probably not have gotten so excited about it. Nevertheless, invertebrate trace fossils in Alabama also record valuable clues to the behavior of ancient organisms. The invertebrate trace fossils, like the vertebrate tracks, tell us about creatures that left no other clues to their existence. They complement what can be deduced from the vertebrate trace fossils and allow a more complete picture of coal-age life.

INSECTS

The ichnogenus *Stiaria* was first described by Smith (1909) and consists of a trackway with a central drag mark and repeated sets of paired, simple, linear or curvilinear depressions, three on each side of the

FIGURE 16.1 Examples of the ichnogenus *Stiaria*, attributed to a jumping wingless (monuran) insect. These traces were made when the insects used only very short jumps to essentially walk or crawl like many small birds do. The insects were hexapodal (six legged) and may have been gregarious (gathered in groups). (*A*) UCM 1748; (*B*) UCM 1758; (*C*) UCM 1749. *B* shows both surface tracks (with tail drag) and undertracks (without tail drag). The latter would have been impressed later, after the other tracks had been buried in sediment. The trackways in *C* have broad medial impressions (thorax drag) and appear to have been made in wetter mud than those in *B*.

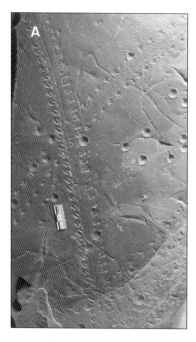

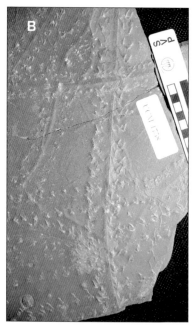

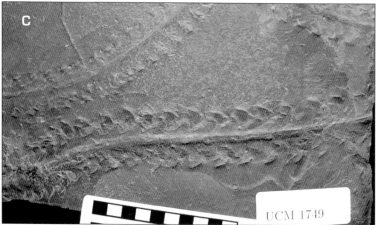

central drag. The groups of three prints on either side are opposite one another (approximately "in phase"), as if the animal progressed by making short hops.

Stiaria is the most common invertebrate trackway found at the Minkin Site. The examples shown in figure 16.1*A–C* illustrate the significant morphological variations in the genus, in some cases showing a body drag rather than a tail drag. Like the vertebrate trackways, most Minkin Site *Stiaria* are undertracks, and some retain only two pairs of depressions per set and lack the central drag. At this time, the *Stiaria* traces at the Minkin Site have not been assigned to a specific ichnospecies.

Stiaria is problematic at a site where undertracks dominate, because it is possible for two completely unrelated ichnogenera to have very similar undertracks. The horseshoe crab trace fossil *Kouphichnium* has also been identified at the Minkin Site and is normally easily distinguished from *Stiaria*, but in deeper undertracks, *Kouphichnium* can resemble undertracks of *Stiaria*. In fact, in the Union Chapel Monograph, almost all the traces originally recognized as *Kouphichnium* are now interpreted as *Stiaria*. One reliable clue for shallower undertracks is that *Stiaria* tracks are simple; they lack the bifurcating ends that are characteristic of *Kouphichnium*. However, these bifurcations tend to vanish from deeper undertracks. Minter and Braddy (2006) showed that *Stiaria* traces were likely made by wingless hexapodal (six-legged) monuran insects (Buatois et al. 1998) that were capable of locomotion by both walking and jumping. The central drag mark would be from the animal's tail, also known as the terminal filament.

Figure 16.2 shows an artist's rendition of a monuran insect making *Stiaria*. The model for the trackway is UCM 1748 (fig. 16.1*A*), and the outline of the insect is based on Minter and Braddy (2006). The three grooves on either side of the central drag mark are from the three different legs on each side of the insect's body. The inner grooves parallel to the central drag mark were likely made by the forelegs of the insect, while the two outer grooves were made by the legs on the thorax (main body section), which extend to slightly different distances.

An atlas of Minkin Site invertebrate trace fossils (Buta et al. 2005) illustrates the apparent gregarious nature of the *Stiaria* tracemakers, which like many invertebrates evidently populated the substrate in large numbers. A single large rock (the "Big Rock," previously mentioned in chapter 3) yielded multiple horizons densely covered in *Stiaria* trackways. No vertebrate slab found at the Minkin Site comes close to showing such large numbers of overlapping trackways. All the specimens shown in figure 16.1*A–C* are from this rock. Although it looks like the tracemakers were walking all over each other in UCM 1758 (fig. 16.1*B*), this was not necessarily the case. Perhaps there were multiple insects at the same time, but a dense track horizon could mean that the area was well traveled over a period of time. A given horizon could include surface tracks from one time and undertracks *from a later time.*

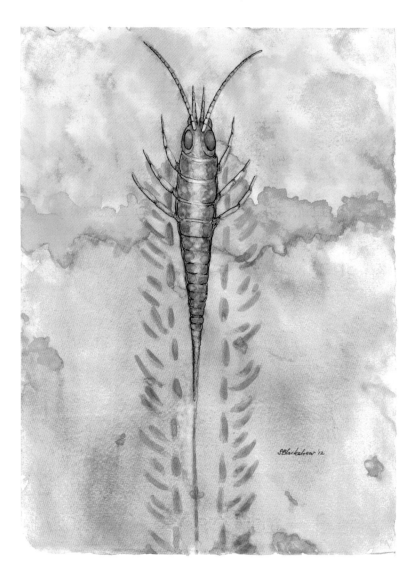

Figure 16.2. Artist's conception of a wingless insect leaving *Stiaria* tracks on the mudflat.

The Big Rock does illustrate something important about the makers of *Stiaria*: their role in the ancient food web. The main Minkin Site assemblage represents the creatures that lived on a tidal flat. The wingless insects thought to have made *Stiaria* probably fed on detritus and minute organisms that became available each time the tide went out. There would have effectively been a "feeding frenzy," just as on modern tidal flats. Living relatives—seashore bristletails—do the same thing. This large number of insects would attract hungry tetrapods. A significant part of the Pennsylvanian food web is preserved with the plants, insects, and tetrapods (N. Minter, pers. comm. 2010).

Prominent on several of the specimens from the Big Rock are large circular depressions or elevations that are likely the result of gas bubbles (chapter 9). The Big Rock yielded some of the best examples of gas bubbles found at the site. Some of the gas-bubble impressions are more than 1 cm in diameter. UCM 1748 (fig. 16.1*A*) and UCM 1749 (fig. 16.1*C*) also show the larval insect burrow *Treptichnus apsorum*, which is discussed further below.

Stiaria is clearly an important part of the trace fossil assemblage of the Minkin Site. Not only is the ichnogenus abundant, but some of the finest and most regular trackways found at the site are of this type. The Big Rock provided many excellent examples.

Tonganoxichnus robledoensis (Mángano et al. 1997; Braddy and Briggs 2002) is named for the Robledo Mountains of New Mexico, famed for vertebrate trackways and invertebrate traces from the Permian Period (chapter 10). *T. robledoensis* is a jumping trace that appears in the form of connected or nearly connected repetitive, complex patterns. Several examples from the Minkin Site are shown in figure 16.3*A–C*; figure 8.11 shows an excellent example on the large slab UCM 1060. The ichnospecies is not common at the site, occurring on only 59 of 2,201 specimens, or 2.7 percent of the online database. *T. robledoensis* is about the same width as *Stiaria*, and they commonly occur together (fig. 16.3*C*). Both trace fossils seem to have been made by hexapodal arthropods of the same size. In fact, several examples have been found of *Stiaria* turning into *T. robledoensis* and vice versa. Three (UCM 1349, 1410, and 5000b) are shown in figure 16.4*A–C*. The first two are simply typical examples in which a *Stiaria* trackway appears to end abruptly in

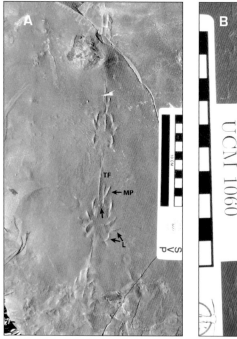

FIGURE 16.3*A–B* Examples of *Tonganoxichnus robledoensis*, attributed to a wingless jumping monuran insect, the same kind of insect that left the *Stiaria* traces. (*A*) UCM 2466; (*B*) UCM 1060. The small repetitive patterns in *A* are places where the animal landed with its six legs oriented forward (up in the picture), leaving an impression of its feeding appendages (maxillary palps, MP) and mouth (mandibles, middle arrow). Then, the creature reversed its hind legs (L) and set its tail (or terminal filament, TF) down in the sediment to jump to the next location. The terminal filament is the linear feature that roughly joins different jump traces. In *B*, which shows the same slab as figure 8.11, the jump patterns are more complex because the creature walked a little before its next jump.

a close pair of indentations. We can see that UCM 1410 becomes a partial *T. robledoensis* trace directly in the line of movement. UCM 5000b, on the other hand, dramatically connects the two types of traces in a single long trackway. The left part of the top panel looks very similar to the upper right trace in UCM 1349, while the right side of the lower frame looks like UCM 1410. The patterns must be undertracks because the terminal filament is less distinct in the *Stiaria* part than it is in the traces shown in figure 16.1*A–C*, except in areas where the animal was poised to jump. From this specimen, there can be no doubt that the two types of trace fossils, *Stiaria* and *T. robledoensis*, were made by the same animal, likely a monuran insect (Minter and Braddy 2006).

Jumping traces of hexapodal insects show complex shapes because of both takeoff and landing behavior. In *Stiaria* traces, the insect is mostly crawling, and its walking grooves curve in the direction of its locomotion. The *T. robledoensis* traces, in contrast, show appendages that curve backward (as in fig. 16.3*A*). This involves the repositioning of the main thoracic legs for the next jump. According to Mángano et al. (1997), the

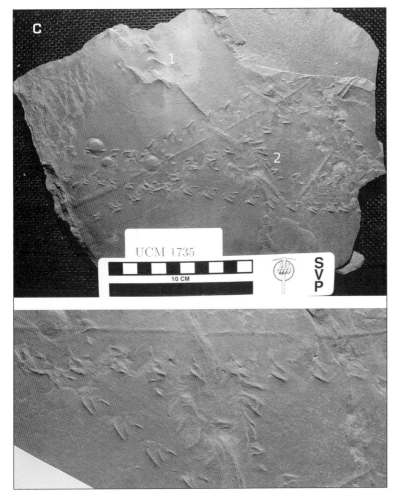

FIGURE 16.3C
Continued examples
of *Tonganoxichnus
robledoensis,* attributed
to a wingless jumping
monuran insect, the same
kind of insect that left the
Stiaria traces. UCM 1735.
The numbers point to
two *T. robledoensis* traces
mixed with three *Stiaria*
trackways; the bottom
frame shows a close-up
of trace 2.

maker of *Tonganoxichnus* traces could jump backward or forward, using
its terminal filament as a launching mechanism. In the prejump position, its tail would be curled up and its rear thoracic legs would be oriented backward. To jump, it would lower its tail and push down, and as
it jumped, it would curl its head toward its thorax. As it landed, its tail
would curve up and its six legs would be oriented forward. Also, impressions would be made of the insect's maxillary palps, or feeding appendages. Then the process would repeat. The examples in figure 16.3*A–C*

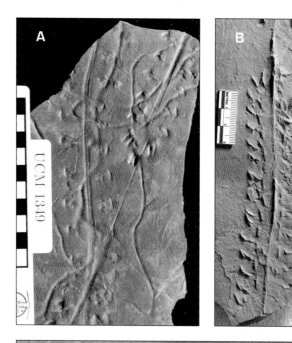

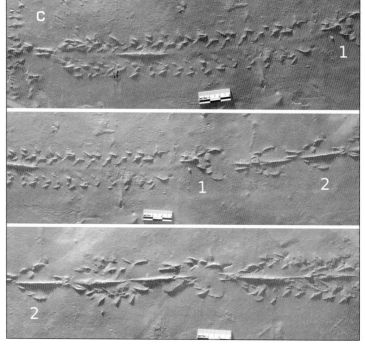

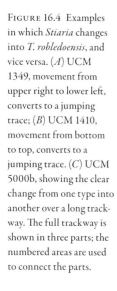

FIGURE 16.4 Examples in which *Stiaria* changes into *T. robledoensis*, and vice versa. (*A*) UCM 1349, movement from upper right to lower left, converts to a jumping trace; (*B*) UCM 1410, movement from bottom to top, converts to a jumping trace. (*C*) UCM 5000b, showing the clear change from one type into another over a long trackway. The full trackway is shown in three parts; the numbered areas are used to connect the parts.

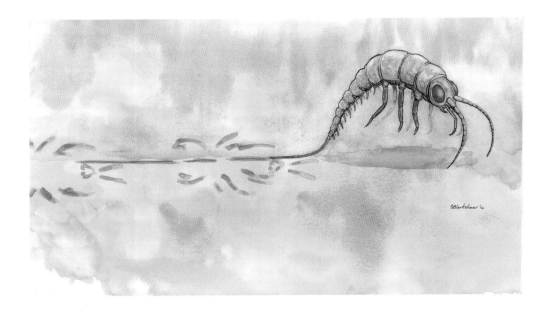

show that Minkin Site monurans generally jumped less than two body lengths. This is different from the makers of the Robledo Mountains specimens, which were smaller and could jump several body lengths. These could also use their *abdomens* as a launching vehicle (Minter and Braddy 2006).

FIGURE 16.5 Artist's conception of a wingless insect traveling by jumping and leaving *T. robledoensis* on a mudflat.

Another excellent discussion of *Tonganoxichnus* appears in Minter and Braddy (2009). Like *Stiaria*, this is one of the most beautiful of invertebrate Minkin Site traces. Figure 16.5 shows a monuran insect making a sequence of *Tonganoxichnus* traces.

XIPHOSURANS (HORSESHOE CRABS)

Horseshoe crabs have a long history in the geologic record (Twenhofel and Shrock 1935). Like insects, horseshoe crabs are arthropods with a hard external skeleton and multiple appendages for locomotion and feeding. They get their name from the shape of their carapace (main part of the exoskeleton), which looks like a horseshoe. Their abdominal section can bend away from the carapace and terminates in a stiff tail-like appendage called a telson. This assists the animal in moving through mud and in reorienting itself if it gets turned over. In the abdomen is a complex gill structure, and five pairs of legs for walking. Horseshoe crabs have mouth appendages called chelicerae. When they move, they look like walking army helmets. Xiphosurans, the scientific name for horseshoe crabs, comprise a large group of species, most of them extinct. The only surviving members of the order Xiphosura belong to a subgroup called the limulids. As fossils, xiphosurans have been traced back to the Ordovician Period, 450 million years ago.[1] Because the four living species so closely resemble their distant ancestors, horseshoe crabs are sometimes referred to as "living fossils."

Although horseshoe crabs are marine animals, they can live in estuaries. When they breed, they come ashore to lay their eggs in sand. This is probably how horseshoe crab trace fossils ended up at the Minkin Site. Trace fossils made by horseshoe crabs are called *Kouphichnium* (Nopsca 1923) and in the best preserved cases are characterized by *Y*-shaped footprints laid out in complex patterns. Figure 8.10 shows an excellent example; others are shown in figure 16.6. Hundreds of specimens from the Minkin Site were initially called *Kouphichnium*. However, Nic Minter examined these and concluded that many were misidentified. At this point, there is no telling how many different kinds of multilegged creatures were crawling around on the mudflats 313 million years ago. For now, we follow Minter, who notes: "*Kouphichnium* is a trackway that has groups of five tracks on either side of a linear medial impression. The tracks on either side have opposite symmetry [i.e., the equivalent tracks on either side line up with one another and are 'in phase'] and the inner four tracks on either side are simple and linear in form with occasional bifurcating [split] terminations. The outer tracks on either side have a different morphology with a central region and then several imprints around this that make them superficially look like a tetrapod footprint" (N. Minter, pers. comm. 2010). In a classic study, Ken Caster (1938) provided a detailed description and analysis of trackways that are now called *Kouphichnium* and compared them to trackways made by living horseshoe crabs in fresh, brackish (semisaline), and saltwater habitats. He also demonstrated that they were not of vertebrate origin.

Horseshoe crabs are unusual in that they have different kinds of legs. Reptiles and amphibians have four feet that are quite similar to one another, except the front feet are usually smaller. Many arthropods, such

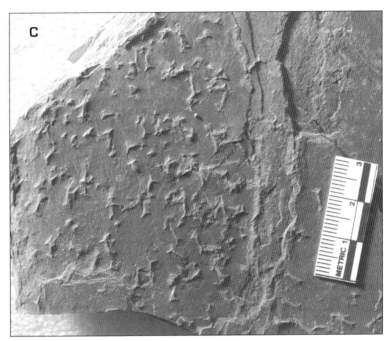

Figure 16.6 Examples of the ichnogenus *Kouphichnium*, attributed to xiphosurans, primitive horseshoe crabs. (*A*) UCM 437; (*B*) UCM 1505; (*C*) UCM 1268; (*D*) UCM 117. The hallmark of this ichnofossil is complex patterns caused by the five pairs of legs, most of which are biramous (split, like the letter *Y*) at the tips. *A* shows these *Y* shapes clearly, as does *B*. *C* contains a large number of disorganized *Y*s and may have been a congregating area. *D* is complex but the *Y* shapes are clear; this is another common form of *Kouphichnium*.

as trilobites and millipedes, have feet that all look about the same in any given species. Horseshoe crabs, however, have four or five pairs of walking feet that end with small pincers known as chelae. The *Y* shapes in their trackways are the result of the biramous nature of the appendages, which have two tips. The rear legs on a horseshoe crab are different: they are tipped with spiny paddles, making them easy to recognize. Sea

FIGURE 16.7 Artist's conception of a small horseshoe crab leaving a *Kouphichnium* trail on the mudflat, based on UCM 437.

scorpions (eurypterids) were related to horseshoe crabs and were also heteropodous (had more than one kind of foot). It takes detailed detective work, and body-fossil information, to tell exactly what kind of creature made *Kouphichnium* and related traces.

The *Y*-shaped appendages at the tips of the footprints of the tracemaker can be seen in figures 8.10 and 16.6. The lack of a telson impression in general suggests that all but those in figure 8.10*A* are undertracks. Telson drag may be present in UCM 117 (fig. 16.6*D*) but is offset to one band of the trackway. In this case, the telson appears to have been applied intermittently to the mud, rather than continuously (Caster 1938).

UCM 437 (fig. 16.6*A*) shows well-defined *Y*-shaped prints in complex groups but no trace of a telson drag. The trackway strongly resembles one shown by Caster (1938), in which the intermittent application of the telson leaves some sections with no telson impression. UCM 437 may be shallow undertracks.

Specimens UCM 1505 (fig. 16.6*B*) and UCM 1071 (fig. 8.10*B*) show parallel sets of isolated *Y*s and are definitely undertracks, as demonstrated directly by UCM 1071, which appears on the underside of the trace shown in figure 8.10*A* (UCM 1070).

Figure 16.7 shows an artist's rendition of a horseshoe crab walking across the mudflat and leaving the trail of footprints in UCM 437. Based on specimens of *Kouphichnium* from the Minkin Site, and information from Seilacher (2007), we deduce that the tracemakers had carapaces 4 to 9 cm (2 to 4 in) across. The tracemakers apparently belonged to a species of horseshoe crab much smaller than the familiar Atlantic horseshoe crab, *Limulus polyphemus*, although the carapace of adults of one living species, *Carcinoscorpius rotundicauda*, is only 10 to 12 cm (4 to 5 in) across.

MYRIAPODS

A myriapod is an arthropod with a multitude of legs: a centipede or millipede. Such arthropods have a distinct head, eyes, jaws, a single pair of antennae, a wormlike body, and multiple body segments from which one or two pairs of leglike appendages emerge (Twenhofel and Shrock 1935). Fossil myriapods have been identified at least as far back as the late Silurian, more than 420 million years ago.[2]

Figure 16.8. Examples of *Diplichnites gouldi*, attributed to a myriapod (many-legged animal), likely a millipede. (*A*) UCM 666; (*B*) UCM 1267; (*C*) unnumbered specimen at McWane Science Center (collected by Ron Buta). Also see figure 8.12 for an exceptional example. This and *A* are most typical of the ichnospecies in showing pairs of closely spaced dimples, suggesting a creature with pairs of pointed legs on each segment of the body. In *B*, because of the way the animal walked, the leg impressions left doubled short grooves that appear to consist of closely spaced dimples, rather than single dimples. *C* is a very rare case in which a *D. gouldi* trackway abruptly ends (arrow), in much the same way as the *Stiaria* traces in figure 16.4*A* and *B* end. What happened to the creature is uncertain. The lower image is a close-up of the termination point. Unlike the other specimens shown, this one has a clear body-drag impression.

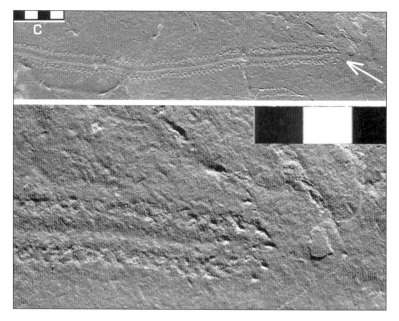

Trace fossils made by myriapods have been found at the Minkin Site. The ichnospecies recognized is called *Diplichnites gouldi* (Gevers et al. 1971) and commonly appears as meandering parallel rows of closely spaced, very small dot-like imprints (figs. 8.12 and 16.8*A*). Typically each side of the trackway consists of a double row of dots. Some imprints are ellipsoidal or transversely directed scratches (fig. 16.8*B*) (Lucas and Lerner 2005). These distinctive trackways are uncommon at the Minkin Site, occurring on 19 of the 2,201 slabs (less than 1 percent) of the online database. A few of these examples occur with other trace fossils, notably *Treptichnus apsorum* and even *Attenosaurus subulensis* (UCM 2528; Buta et al. 2005, plates 35*A*, 35*B*, and 110*C*). Good illustrations of *D. gouldi* from Kansas (the state) were published by Buatois et al. (1998). *Diplichnites* was reproduced quite well in an experimental study using a living giant millipede (Davis, Minter, and Braddy 2007).

Some authors have ascribed *Diplichnites* to other kinds of arthropods, such as trilobites. In opposition to this view is that myriapods have many pairs of legs and do not taper much from head to tail. Most Carboniferous trilobites had eight to ten pairs of legs, and they tapered sharply. Their rear legs would have been set down in the sediment close together compared to their front legs, which would have yielded traces resembling nested *V*s or multiple overlapping rows of foot impressions. Most trilobites would have been hard pressed to make the simple double rows of imprints that comprise *Diplichnites*. Trilobite walking traces are more commonly assigned to *Petalichnus* than to *Diplichnites* (for example, Rindsberg 1990).

A unique example of *D. gouldi* from the Minkin Site is shown in figure 16.8*C*. This case appears to terminate abruptly in a rounded dot pattern that is similar to abruptly terminating *Stiaria* traces (fig. 16.4*A*, *B*). There is no evidence on the rock that something picked the creature up, nor does it appear to have burrowed into the sediment. Millipedes, unlike monurans, are incapable of long-distance jumping. Could a giant dragonfly lift a millipede, and do it so quickly that there would be no sign of a struggle? Or does this specimen record a millipede walking onto drier ground where its feet left no mark?

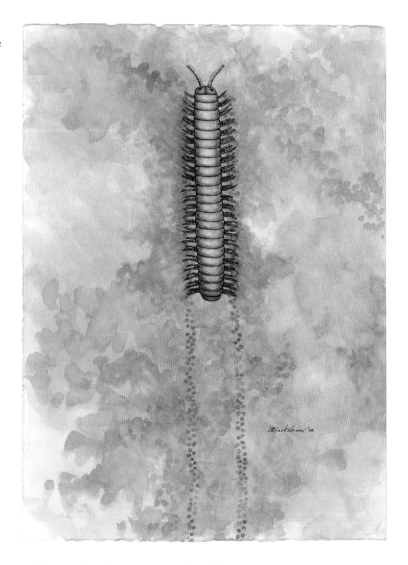

FIGURE 16.9 Artist's conception of a millipede leaving paired trails of dimples on the mudflat, based on UCM 666.

Figure 16.9 shows an artist's rendition of a Minkin Site millipede making the trace seen in UCM 666 (fig. 16.8*A*). The trace shows how wide the animal was, but not how long.

LARVAL INSECT TRACES

Some animals feed by burrowing, and since burrowing is easier in soft ground, a mudflat should be a good place to find burrows. Because burrows can go farther below the surface than undertracks, they have a greater preservation potential than trackways (Seilacher 2007). However, the common burrows at the Minkin Site formed only millimeters below the surface.

The two main types of burrows identified are *Arenicolites longistriatus* and *Treptichnus apsorum* (Rindsberg and Kopaska-Merkel 2005).[3] Both are ubiquitous at the Minkin Site, and both are considered new ichnospecies. Together they are interpreted as different expressions of burrowing by insect larvae (as explained below). However, they look very different. *A. longistriatus* consists of shallow *U*-shaped burrows (fig. 16.10*A*), and *T. apsorum* (fig. 16.10*B*, *C*) brings together several varieties of zigzag horizontal burrows. These trace fossils are the two most common at the site, commonly occur together, and are shown by the characteristics of transitional forms to have been made by the same animals.

Invertebrate trace-fossil species are named by reference to their morphology, not according to what type of animal is thought to have made them, as is more often the case with vertebrate tracks. This is because interpretations can be modified in the light of new information, but traces do not change. If trace fossils are classified on the basis of what made the trace, then for consistency, when interpretations change, names would also need to change. Sometimes this has happened, leaving tracks and traces with inappropriate names. This is why it is best (safest) to classify traces on the basis of morphology.

A. longistriatus and *T. apsorum* are associated with nearly all other trace fossils found at the Minkin Site, including tetrapod trackways. Other species of *Arenicolites* and *Treptichnus* from other parts of the world are similar to the two described here, but they are not necessarily made by the same organisms. The Minkin Site specimens share certain diagnostic characteristics that strongly indicate that both were made by one organism. For example, *Arenicolites* from the Minkin Site have longitudinal grooves scored into the floors of the burrows (fig. 16.10*A*). These grooves strongly resemble those in burrows made by modern fly larvae (Rindsberg and Kopaska-Merkel 2005). Specimens of *T. apsorum* from the Minkin Site show the same kind of longitudinal grooves. Burrows made by modern fly larvae also resemble the Carboniferous *T. apsorum*

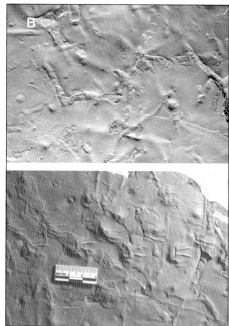

FIGURE 16.10*A* Larval insect burrows, *Arenicolites longistriatus*. *Top*, UCM 2038, large striated specimen, with very small *Treptichnus apsorum* to the right, negative epirelief; *bottom*, UCM 2038, large striated specimens, with small *T. apsorum* and gas-bubble impressions, positive hyporelief.

FIGURE 16.10*B* Multiple larval insect burrows, *Treptichnus apsorum*. *Top*, UCM 2026, positive hyporelief; *bottom*, UCM 448 is similar to UCM 2026, but a tetrapod trackway crosses the area.

in another way: many specimens of *T. apsorum* consist of successive linear segments oriented at distinctly different angles, each segment jutting a short distance past the origin of the next one (arrows in fig. 16.10*C*), just as in burrows made by modern fly larvae (Uchman 2005).

Body fossils of dipterans, the scientific name for flies and their relatives, are known only as far back as the Late Triassic (Evenhuis 2004). A few arthropod body fossils have been found from the Minkin Site, including wings of large dragonflies. The Carboniferous burrows might have been made by early dipterans whose fossils have not yet been found, or by other arthropods of similar body plan and behavior. The diameters of burrows give us at least a range for the width of the creatures that made them. The longitudinal scratches indicate they had sharp-tipped legs. Because many specimens of *T. apsorum* consist of connected line segments, Rindsberg and Kopaska-Merkel (2005) inferred that, for a given burrow, the maker was no longer than the length of the shortest burrow segment. All these characteristics are consistent with those of fly larvae, or other arthropods with a similar form.

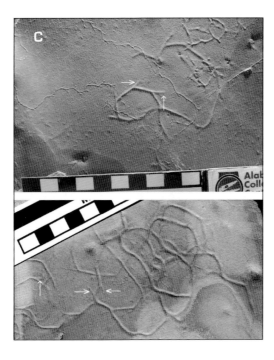

FIGURE 16.10C Larval insect burrows, *Treptichnus apsorum*, positive hyporelief. *Top*, UCM 5031; *bottom*, UCM 417. Arrows point to characteristic burrow extensions.

INVERTEBRATE RESTING TRACES

A resting trace is a burrow made by an animal when it digs in through the muddy surface to protect itself from predators, to feed, or for some other reason. As Seilacher (2007) noted, the term is a misnomer because it assigns to a burrow a particular purpose, "resting," that may differ from its actual purpose. He noted that an animal first uses its legs or appendages to remove the sand under its body and then smooths the surface around it in order to hide. When it eventually leaves its hiding place, the burrow records a body-like outline whose shape depends partly on how the animal left the burrow. Leg impressions may also be impressed in the burrow. If this is covered later by sediment, it can be preserved as a *resting trace fossil*.

Two types of resting traces are found at the Minkin Site: one defined by leg impressions (*Arborichnus repetitus*) and the other defined by a body outline (*Rusophycus*). *A. repetitus* (Romano and Meléndez 1985) is a trace fossil that appears as paired grooves (impressions) or ridges (counterimpressions) flanking an oval central disturbed area.

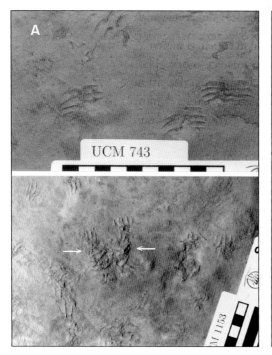

10 CM

FIGURE 16.11*A Arbor-ichnus repetitus*, an invertebrate resting or feeding trace, positive hyporelief. *Top*, UCM 743; *bottom*, UCM 1153; the arrows point to two traces where the animals appear to have moved in the substrate.

FIGURE 16.11*B Ar-borichnus repetitus*, an invertebrate resting or feeding trace. *Top*, UCM 2387, positive hyporelief; *bottom*, UCM 2539, negative epirelief.

Six examples from the Minkin Site are shown in figure 16.11*A–C*. The most well-preserved specimens display four complete pairs of grooves, which are straight to strongly curved. Curved specimens are assumed by convention to arc out and toward the "rear" of the animal. In the central area where the pairs appear to come together, the grooves can again arc to the rear. Minkin Site trace fossils assigned to this ichnospecies were briefly described by Lucas and Lerner (2005; mistakenly called *A. repetita*) and photographed by Buta et al. (2005).

A. repetitus is interpreted as an invertebrate resting or feeding trace. The makers appear to have had four or five pairs of legs. The central disturbed area suggests that the body was pressed into the sediment, likely during push-off. Another possible explanation for the central disturbed area is feeding at the sediment surface, but it is difficult to see how the appendages around the mouth could have reached as far back as the last pair of legs. The trace was made by an arthropod with at least eight legs, perhaps a decapod crustacean (with ten legs). The maker is not known, but it was probably not a horseshoe crab.[4] As shown previously, the

FIGURE 16.11C *Arborichnus repetitus*, an invertebrate resting or feeding trace, positive hyporelief. *Top*, UCM 2449, traces made by animals hopping about one body length forward or backward; *bottom*, UCM 3780, "extra" leg impressions made by animal shifting in place.

"tails" of horseshoe crabs usually leave drag marks, their front appendages are pincers and commonly leave *Y*-shaped grooves, and their back legs are shaped like paddles.

Specimens assigned to *A. repetitus* are found on thick, solid sandstone slabs. They come from higher in the stratigraphic section than the *Cincosaurus* beds, in rock layers that may have been deposited when sea levels were higher and a mudflat no longer existed in the area. This could explain why *A. repetitus* occurs with few other trace fossils. In his stratigraphic analysis of the Minkin Site highwall, J. Pashin (chapter 9) placed *Arborichnus* near the top of the section, where the layers could be thousands of years younger than the *Cincosaurus* beds.[5] Whatever the conditions under which the *A. repetitus* tracemakers lived, almost nothing else was there to make a lasting impression on the sediment surface.

A. repetitus is also one of the few ichnospecies at the Minkin Site that can actually be collected in situ. The traces are impressed in rock layers that are close to top of the highwall, which can be accessed on the west side of the site.

UCM 743 (fig. 16.11*A*, top) shows the typical way *A. repetitus* specimens are found at the Minkin Site: as similar-sized individual patterns scattered like Fourth of July fireworks over a large slab. These tend to be undertraces and the tracemakers appear to have been still when the traces were preserved. UCM 1153 (fig. 16.11*A*, bottom) also displays scattered individuals, but the more complex shapes of some (especially the two flanked by arrows near the center-left) suggest movement that may involve the actual burrowing process. The appearance of so many *A. repetitus* traces on the same horizon of these large slabs suggests that the organisms commonly clustered together.

UCM 2449 (fig. 16.11*C*) shows two distinctly different *A. repetitus* traces. The one on the right is a relatively normal specimen with four major pairs of grooves (ridges in this case, since the specimen is a counterimpression). The one on the left shows many more than four pairs and appears to have moved a short distance. This suggests that the grooves of *A. repetitus* are indeed leg impressions.

Although rare in the online database (twenty-six slabs with probably more than one hundred traces), *A. repetitus* is actually one of the most common trace fossils found at the Minkin Site. The ichnogenus's underrepresentation in the database is mostly a result of collector bias; many specimens are on slabs too large to lift and too thick to break apart.

Arborichnus formed when the tracemakers settled on the tops of high-energy sand beds when the sand was not moving. In other words, the sand layers record high-energy events, perhaps major storms, and the trace fossils on their upper surfaces record subsequent quiet times.

The second type of resting trace found at the Minkin Site is called *Rusophycus* (Hall 1852). The burrows have a roughly oval shape and two lobes with paired, transverse scratches ranging from a few to more than a dozen pairs. Generally found in positive relief on the bottoms of rock beds, *Rusophycus* is an arthropod resting trace, almost exclusively Paleozoic in age and made mostly by trilobites. Two examples are shown in figure 16.12*A*. Pennsylvanian trilobites had between eight and

FIGURE 16.12*A Rusophycus* isp. (trilobite resting traces), positive hypore-lief. *Top*, UCM 3781; *bottom*, UCM 1548.

seventeen pairs of legs; most had eight to ten pairs (Harrington et al. 1959). The two specimens of *Rusophycus* shown have seventeen pairs of transverse scratches, which is consistent with a trilobite origin. They are also well within the size range of known Pennsylvanian trilobites. However, there are good reasons to think that some *Rusophycus* were not made by trilobites. First, some are younger than known trilobites. Trilobites died out at the end of the Paleozoic Era in the greatest mass extinction event the world has ever known, when 95 percent of known species disappeared from the fossil record. Second, some *Rusophycus* are associated with other trace fossils and body fossils that otherwise appear to form only in freshwater or on land; others are found in rock units

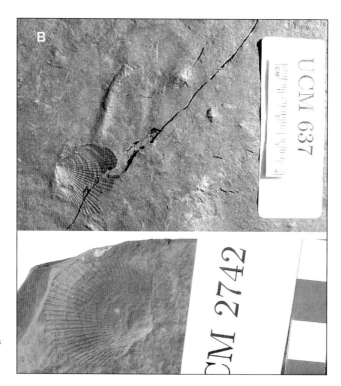

FIGURE 16.12*B* Pelecypod shells, likely the genus *Dunbarella. Top,* UCM 637; *bottom,* UCM 2742.

that are inferred to be freshwater deposits. Third, some ancient crustaceans may have been able to make a *Rusophycus*-like burrow.

Trilobites, responsible for most occurrences of *Rusophycus*, were entirely marine, but other arthropods, such as spiders and scorpions, invaded freshwater and land before trilobites went extinct. Some of these adventurers may have been able to make a trace fossil like *Rusophycus*, so this resting trace is not an unequivocal indicator of a marine environment. Any compact arthropod with many pairs of similar legs can make *Rusophycus* (Collette, Hagadorn, and Lacelle 2010; Donovan 2010). This arthropod body plan is not common today, nor has it been for many millions of years. Most modern arthropods have specialized limbs that have evolved to serve the creatures well in a variety of ways.

Rusophycus at the Minkin Site are found on the undersides of hard, thick sandstone beds that came from a stratigraphic layer higher in the highwall than the *Cincosaurus* beds. Minkin Site *Rusophycus* are very rare, and (like *Arborichnus repetitus*) are likely many thousands of years younger than these beds. They are not associated with any of the estuarine trace fossils from the *Cincosaurus* beds, such as horseshoe crab traces and fish fin traces, or with *A. repetitus*. The small size (and therefore shallow depth) probably limited the preservation potential of these burrows. The *Rusophycus*-bearing stratigraphic interval also includes impressions of shells that are likely pectinid bivalve marine mollusks (scallops) of the genus *Dunbarella* (M. Lockley, pers. comm. 2012), known to have lived during the Carboniferous. Thus, *Rusophycus* were probably made by trilobites and, along with these shells (fig. 16.12*B*), document marine interludes during Pottsville time at the Minkin Site.

OTHER MINKIN SITE INVERTEBRATE TRACE FOSSILS

Lucas and Lerner (2005) described a few other invertebrate ichnofossils that have been found at the Minkin Site. Slab UCM 1650 is densely covered with horizontal burrows thought to be made by insect larvae and nematodes (roundworms). The smaller burrows were assigned to *Cochlichnus* and the larger to *Palaeophycus*. In addition, Lucas and Lerner identified a slab with repeated crescentic, elevated impressions thought to be produced by juvenile horseshoe crabs, which they assigned to *Selenichnites*. Another slab showed a horizontal burrow with a narrow central furrow and chevron-like ridges. This they assigned to *Protovirgularia*, thought to have been made by a primitive bivalve.

There is also strong evidence in the database for trackways produced by terrestrial scorpions (Eiseman and Charney 2010; P. Atkinson, pers. comm. 2011). These were misclassified in the Union Chapel Monograph as *Kouphichnium*, reflecting another case of ambiguity in trackway morphology. In the online database, the following specimens include possible scorpion traces: UCM 73, 802, 959, 1341, 1368–69, 1376–78, 1381, 1384–86, 1390, and 1392. UCM 1377 is shown in figure 16.13*A* and UCM 1368 is shown in figure 16.13*B*. These resemble trackways named *Octopodichnus* and *Paleohelcura* (Seilacher 2007; Eiseman and Charney 2010; P. Atkinson, pers. comm. 2011).

As with the vertebrate traces, some invertebrate traces do not fit neatly into the categories described in this chapter. For example, UCM 331 (fig. 16.14*A*) was the first trackway found at the site (chapter 1),

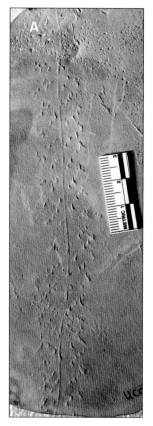
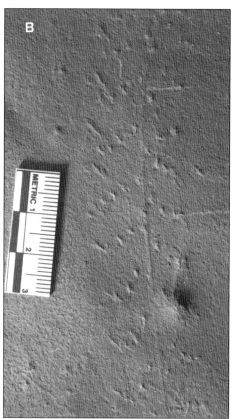

FIGURE 16.13 Possible terrestrial scorpion traces resembling *Octopodichnus* or *Paleohelcura* (Seilacher 2007). The hallmarks of these traces are *V*-shaped patterns of short leg marks and a narrow tail drag. (*A*) UCM 1377; (*B*) UCM 1368.

but its classification has still not been resolved. Haubold et al. (2005a) considered both UCM 331 and UCM 67 (fig. 16.14*B*) to be *C. cobbi* traces made at the original surface, yet both look distinctly invertebrate in nature. UCM 1311 and 1488 (fig. 16.14*C*) appear to be invertebrate surface tracks made in soft mud, but little detail is preserved.

ARTHROPOD BODY FOSSILS

Although body fossils are very rare at the Minkin Site, enough have been found to add a new dimension to our understanding of life in coal-age Alabama. Three insect wings, one trigonotarbid (a spider relative), and two other specimens of unknown nature have been collected by amateurs (fig. 16.15*A–C*). Arthropod exoskeletons are made of chitin, an

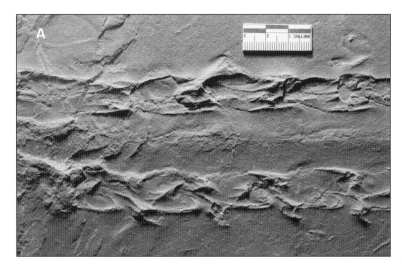

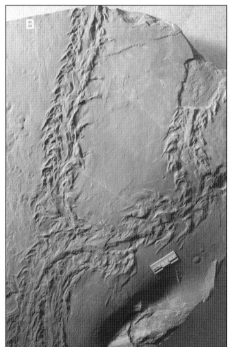

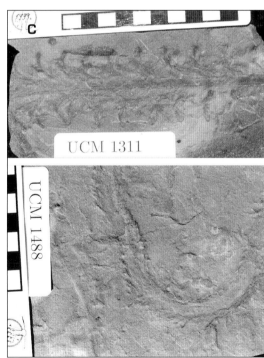

FIGURE 16.14A–C
Four unusual traces of
uncertain interpretation.
(A) UCM 331; (B) UCM
67; (C) UCM 1311 and
UCM 1488. Haubold
et al. (2005a) suggested
that both A and B were
surface tracks of *C. cobbi*.
However, the tracks are
much more complicated
than typical *C. cobbi* traces
and an invertebrate tracemaker is also a possibility
in each case.

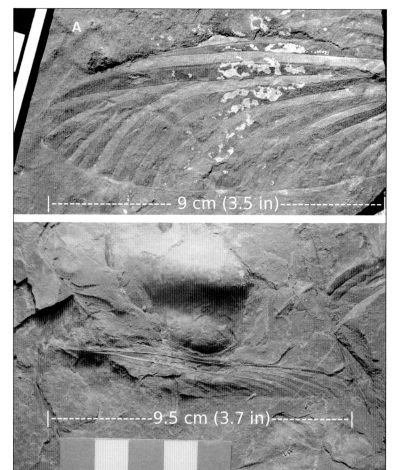

FIGURE 16.15*A* Arthropod body fossils. *Top*, one of the two wings in UCM 1076a, the wing of a giant mayfly, *Anniedarwinia alabamaensis*; see also figure 8.5*A* ; this wing is 9 cm (3.5 in) long and is partly cut off, implying a wingspan of more than 18 cm (7 in); *bottom*, UCM 2369, wing of a giant scorpion fly, *Agaeoleptoptera uniotempla*; this one has a length comparable to that of *Anniedarwinia alabamaensis*.

organic material that is chemically more stable than bone in the acidic environment in and near coal swamps. All the specimens are preserved in laminated shale that seems to be from the *Cincosaurus* beds, and each of the three wing specimens is a different new species. Atkinson (2005) gives a preliminary description of two of these and the trigonotarbid.

Wings have a higher preservation potential than other insect parts because the chitin makes them hard to consume and resistant to decay. Beckemeyer and Engel (2011) studied in detail the wing shown at the top of figure 16.15*A*, one of a pair shown in figure 8.5*A*. The pair, whose discovery we described in chapter 8, is a new species named

Anniedarwinia alabamaensis. Although in figure 8.5*A* the wings look like a forewing and a hind wing, Beckemeyer and Engel argue that they are instead right and left forewings, with one folded over the other. The rightmost wing is the right forewing viewed from below, while the leftmost wing is the left forewing viewed from above. Each wing was estimated to have a total length of 11.5 cm (4.5 in), so if spread out on both sides, together with the animal's body, the wingspan would have probably been at least 25 cm (10 in). Beckemeyer and Engel linked these wings to the insect superorder Ephemeropterida, which means that *A. alabamaensis* was a giant Carboniferous mayfly (Knecht, Engel, and Benner 2011).

The single wing shown at the bottom of figure 16.15*A* is a different type, belonging to the insect superorder Palaeodictyopterida (order Megasecoptera), and which Beckemeyer and Engel named *Agaeoleptoptera uniotempla*, where "uniotempla" means Union Chapel, where the fossil was discovered. The superorder includes primitive insects that could not fold their wings over their abdomens. The order refers to insects that had two pairs of wings of similar size, probably held horizontally as in modern dragonflies.[6] The full wing is 9.5 cm (3.7 in) long, and the wing bearer would have been a fairly large creature compared to its modern relatives (scorpion flies) (Twenhofel and Shrock 1935).

Beckemeyer and Engel also discuss three other new species of wings found in Alabama. *Pharciphyzelus lacefieldi* and *Camptodiapha atkinsoni* are two more Palaeodictyopterids named after J. Lacefield and P. Atkinson, respectively, the collectors of the specimens. *P. lacefieldi* was found at a mine near Eldridge, Alabama, in western Walker County, while *C. atkinsoni* was Prescott Atkinson's third wing discovery at the Minkin Site. The final Alabama wing named was collected in 1988 by J. Hall and K. Gaddy and was on file at the ALMNH. Beckemeyer and Engel assigned this one to the superorder Odonatoptera (order Protodonata), which includes the famous giant dragonflies of the Carboniferous.[7] The preserved length of the wing is 7.5 cm (3 in), and it was named *Oligotypus tuscaloosae*, "tuscaloosae" referring to its discovery in Tuscaloosa County.

The study of these wings from Alabama coal mines has enhanced the significance of the Minkin Site. We now know that the large insects the Pennsylvanian Period is famous for did exist in Alabama. The giant insects were able to thrive during this time because the huge tropical forests of the day raised the level of oxygen in Earth's atmosphere far above what it has been at any other time in Earth's history (35 percent versus 21 percent now; Dudley 1998). Oxygen sharply limits insect size because unlike vertebrates, which have efficient lungs, insects absorb air through their porous exoskeletons. Thus, high levels of oxygen were an advantage to them and allowed them to grow larger than their present-day descendants.

In figure 16.15*B* and *C* are possible small arthropod body fossils. The first is an impression of a trigonotarbid, a kind of arachnid that was like a spider that could not spin webs. UCM 1881 (fig. 16.15*C*, bottom) looks very much like an insect abdomen. This unidentified fossil about 1.5 cm long is found on a small slab with *C. cobbi* tracks, and could be nothing more than a plant fragment. Also unidentified is the strange feature shown in UCM 1272 (fig. 16.15*C*, top); it could be another kind of arachnid impression, or something else altogether.

This chapter has shown the rich invertebrate fauna of the Minkin Paleozoic Footprint Site and brought attention to ichnospecies coming from very different rock layers separated by significant amounts of time. Also, in addition to tetrapods, we now know that while jumping insects,

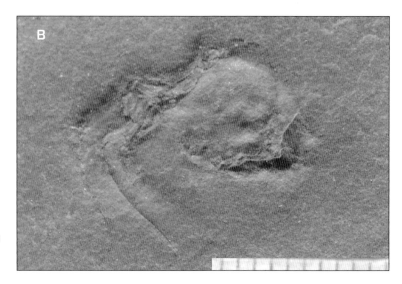

FIGURE 16.15*B* UCM 2281, body fossil of a trigonotarbid, a spiderlike arachnid. Scale bar shows 10 mm.

fly larvae, horseshoe crabs, terrestrial scorpions, millipedes, and trilobites were moving about or burrowing on land or in the estuary, giant ancestors of mayflies, scorpion flies, and dragonflies were buzzing in the air. Some have characterized the Pennsylvanian Period not only as the Age of Coal Swamps or the Age of Amphibians but also as the Age of Arthropods. The trace fossils from UCM are highly consistent with this characterization and, further, provide evidence of the types of invertebrates that were important in the colonization of land at this time.

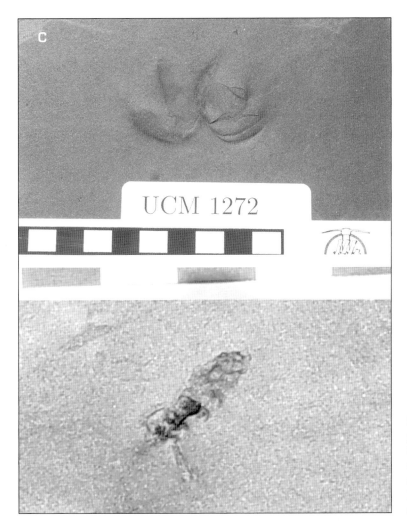

FIGURE 16.15C Arthropod body fossils. *Top*, UCM 1272, a flowerlike impression, perhaps an arachnid; *bottom*, UCM 1881, plant fragment or abdomen of an insect?

17

Fossil Plants

The plant fossils found in Alabama coal mines are truly extraordinary. For no other period of Earth's history is the flora so well preserved. When the area of the Minkin Paleozoic Footprint Site was a dense tropical swamp forest, peat accumulated for thousands of years on the swamp floor. The peat would eventually get buried under tons of sediment and become compressed and heated enough to be changed into coal. The identity of the plants that made the coal would not be preserved. At other times, the forest would not be as dense, and large amounts of peat would not have a chance to accumulate. This would lead to layers rich in well-preserved plant fossils. At still other times, the area was a tidal mudflat, which would lead to layers of rock that would preserve fossil trackways in addition to plants. Plant fossils are very important for understanding the tracemakers' environment, because different environmental conditions (seasonal temperature extremes, amount of rainfall, presence of shore or inland habitat, and so forth) allow different plants to grow (Dilcher and Lott 2005; Dilcher, Lott, and Axsmith 2005). The excellent condition of the material indicates that the plant fossils were not transported far after falling to the forest floor. That is, the plants grew where we find them.

Trees evolved about 364 million years ago (Scheckler 2001) and forests with well-developed canopies were growing by 345 million years ago (Dilcher et al. 2004). By the Westphalian A, the time frame of the Minkin Site (310 to 315 million years ago), forests were well established. Trees had reached their maximum height and multistory forest canopies had developed. Nevertheless, ecosystem complexity continued

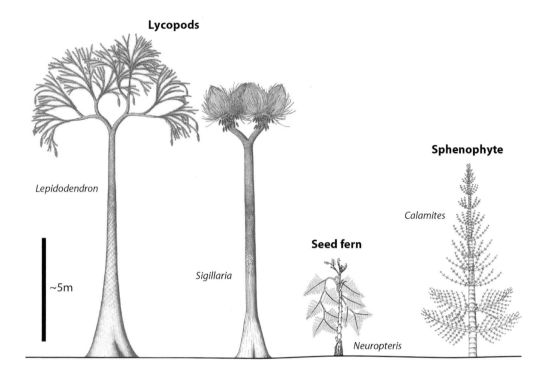

Lycopods

Lepidodendron

~5m

Sigillaria

Seed fern

Neuropteris

Sphenophyte

Calamites

to increase after the Westphalian (Dilcher et al. 2004). Understory plants subsisted on sunlight that made it through the open lycopod canopy (meaning these ancient trees did not have spreading branches like modern trees and so did not provide as much cover) (Scheckler 2001). Many of these plants show physical adaptations for swamp life (Dilcher and Lott 2005; Dilcher, Lott, and Axsmith 2005), such as tissue containing open channels that made it easier for oxygen to reach submerged portions of roots.

Dilcher and Lott (2005) and Dilcher, Lott, and Axsmith (2005) present a photographic survey of all the major Minkin Site plant fossils. Here we describe a few of the plant fossils from their survey. Figure 17.1 shows schematics of the main types we will describe (Pashin 2005) (see also fig. 5.5). The largest are the lycopods, *Lepidodendron*, *Sigillaria*, and *Lepidophloios*, followed by the sphenophytes ("horsetails") and the pteridosperms (seed ferns).

FIGURE 17.1 Schematic reconstructions of common components of the Minkin Site forest.

Arborescent Lycopods

As we described in chapter 5, Carboniferous forests were not like the forests of today. Most trees were naked trunks and did not have multiple branches to form dense canopies like the oaks and other common trees we see in the woods today. Arborescent lycopods, the tallest trees (up to 40 m [130 ft] high), branched once, a few times, or not at all. They were not covered with broad leaves in dense clusters on myriad twigs. Instead, they had long, narrow leaves that were quickly shed. Picture a green telephone pole and you will have a good idea of what they looked like. Their peculiar structure was effective at photosynthesis, which maximized their ability to grow (Dilcher et al. 2004).

Carboniferous forests produced a great deal of coal because of the arborescent lycopods. These huge trees provided most of the biomass of the accumulating peat (Lacefield 2000), which formed because the swamp water was significantly deoxygenated and decay could not keep up with the mass of dying plants. Also, wood-rotting fungi, so prevalent in modern forests, had not yet evolved; these first appeared about the end of the Paleozoic (Dilcher et al. 2004). In their absence, wood was buried and preserved that nowadays is quickly recycled into fungal flesh, which does not normally turn into coal.

For sheer plant fossil beauty, the bark impressions of the arborescent lycopods are unrivaled. The beautiful patterns represent nature at its artistic finest. Shown in figure 17.2*A–E* are several especially well-preserved specimens collected from the Minkin Site. Figure 17.2*A* shows a counterimpression (meaning leaf scars were depressed in the rock, when in life they were elevated cushions) of *Lepidodendron aculeatum*, one of the more common bark species in the Black Warrior Basin. The scars originally adorned the tough outer casing of the plant. The casing surrounded much softer tissue inside, unlike in modern trees, which are woody throughout the trunk. The chute-like leaves of *L. aculeatum* came out of the small diamond-shaped spots in the middle of the larger patterns. In life, the scars would have been elongated vertically, parallel to the tree trunk. As the tree grew, it would shed its "leaves" and leave behind the naked, scarred trunk. Later in life, after achieving a huge size as a green pole, *L. aculeatum* would sprout upper branches (as in fig. 17.1) with chute-like leaves (see below). The foreground of figure 5.4 shows an *L. aculeatum*–like trunk lying on the ground with a large roach walking on it.

Figure 17.2*B* shows a different lycopod called *Lepidodendron obovatum*. Again, the highly detailed pattern is a counterimpression with depressed leaf cushions. As in *L. aculeatum*, the small leaf-scar spots are near one side of the repeated pattern units.

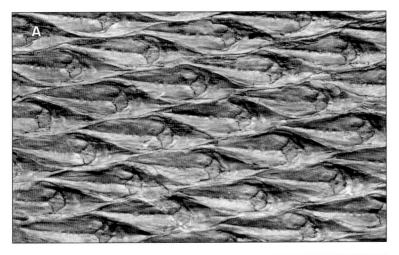

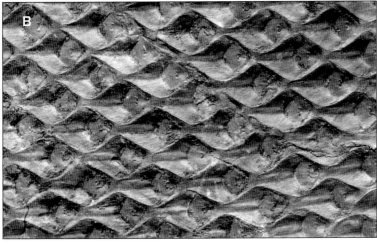

FIGURE 17.2*A–B* Bark impressions of arborescent lycopod trunks, showing the beautiful repetitive patterns that characterized the biggest trees of the Carboniferous Period. (*A*) *Lepidodendron aculeatum*, scar length 38 mm; (*B*) *Lepidodendron obovatum*, scar length 15 mm. These interpretations are based on Dilcher and Lott (2005) and Dilcher, Lott, and Axsmith (2005).

FIGURE 17.2*C–E* Bark impressions of arborescent lycopod trunks, showing the beautiful repetitive patterns that characterized the biggest trees of the Carboniferous Period. (*C*) *Sigillaria elegans*, scar length 8.2 mm; (*D*) young *Lepidophloios laricinus*, scar length 8.5 mm; (*E*) old *Lepidophloios laricinus*, scar length 31.5 mm. These interpretations are based on Dilcher and Lott (2005) and Dilcher, Lott, and Axsmith (2005). *D* and *E* are the same species, but the scars in *E* are much larger and less three-dimensional, which could signify a much larger trunk and greater age for the plant.

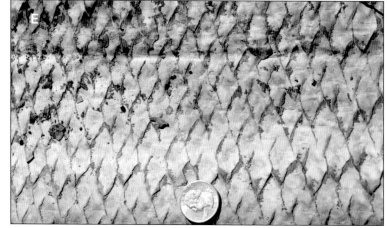

Figure 17.2D shows an extremely well-preserved, highly flattened cast impression of a different bark species, *Lepidophloios laricinus*, in which the elevated scars are visible all the way around the flattened cylinder. The leaf scars are smaller than those of *Lepidodendron aculeatum*, and the bark has fine longitudinal lines not present in *L. aculeatum*. Each diamond-shaped scar has a small spot inside where the leaf came out. This specimen had to be delicately excavated from a large slab and, judging from the small size of the scars and the whole piece, was probably a youthful version of the tree. A more mature example of *Lepidophloios laricinus* has larger scars with little relief (fig. 17.2E). In life, the large scars were elongated horizontally and looked more like fish scales than did those of the other species shown.

Sigillaria elegans was an arborescent lycopod with well-separated hexagonal leaf scars. An excellent example is shown in figure 17.2E. *Sigillaria* was a smaller tree than *Lepidodendron* but had a similar appearance in that most of the branches were at the top of the tree (fig. 17.1).

This sampling of bark impressions from the Minkin Site shows a quality of preservation and diversity superior to that typically found at coal mines in Walker County. The bark, however, is only one part of these trees. Fossils of the upper branches are also found (fig. 17.3, top). These are known as *Lepidodendron lycopodioides* and would have been at the very top of the tree, hanging over like "hair" (as in fig. 17.1). The branches were dichotomous, meaning they split into two parts near the ends. The two branches near the top of figure 17.3 are members of such a pair. This picture also shows the "leaves" of the *Lepidodendron* tree. These are the curved, thin "chutes" of grasslike or needlelike protrusions clustered along the branches, which look very similar to those of the modern plant known as *Lycopodium*, a genus of club moss (fig. 17.3, bottom). *Lycopodium* is little more than underbrush and may creep along the ground. Although the scale trees are sometimes characterized as "giant club mosses," *Lycopodium* is likely only distantly related to them (Lacefield 2000 and pers. communication, 2015).

FIGURE 17.3 *Top, Lepidodendron lycopodioides,* fossils of the upper branches of the *Lepidodendron* genus of scale trees; *bottom, Lycopodium* (green plants), a modern genus that resembles the branches of the ancient scale trees and is likely related to them.

Figure 17.4 shows a cone of the *Lepidodendron* tree, assigned to the genus *Lepidostrobus.* This is the reproductive organ of the giant scale trees. The separate pieces are called sporophylls and include sporangia, or spore cases. When found separately, spore cases are assigned to the genus *Lepidostrobophyllum*; an example is shown in fig. 17.6*E.* The trees were vascular plants (bringing in nutrients and water through a vein-like system) that reproduced by spores rather than seeds.

FIGURE 17.4 *Lepidostrobus*, a cone of the giant scale trees. The pieces are sporophylls, or spore-bearing leaves containing sporangia (spore cases).

SEED FERNS

Seed ferns (also called pteridosperms) were a remarkably sophisticated component of Carboniferous forests. After all, reproduction by seeds, which resist drying, helped plants conquer terrestrial environments. Several species of seed fern have been found at the Minkin Site. Seed ferns have three associated types of fossils: individual fronds (leaves,

FIGURE 17.5 UCM-P 002, the frond of a seed fern, *Neuralethopteris biformis*. The dark material that defines many of the pinnules is the carbonized remains of the original plant. Fine details are preserved in most of these pinnules.

FIGURE 17.6 Seed ferns, cones, and other plant fossils from the Minkin Site: (*A*) *Neuralethopteris biformis* (seed fern); (*B*) *Trigonocarpus ampulliforme* (seed case); (*C*) *Whittleseya elegans* (pollen organ); (*D*) *Lepidostrobus* sp. (cone of arborescent lycopod); (*E*) *Lepidostrobophyllum* cf. *majus*; (*F*) *Neuralethopteris pocahontas* (seed fern); (*G*) *Syringodendron* sp. (decorticated bark of arborescent lycopod); (*H*) *Calamites undulatus* (pith cast of trunk of horsetail); (*I*) *Calamites goepperti* (pith cast); (*J*) *Lyginopteris hoeninghausi* (seed fern).

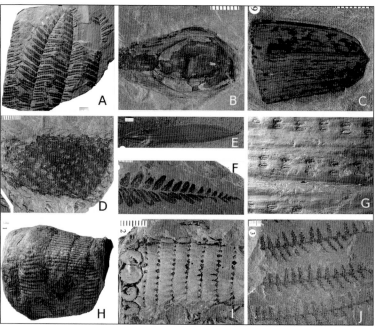

such as figs. 17.5 and 17.6*A*, the common *Neuralethopteris biformis*) and pinnules (petals); seeds (for example, *Trigonocarpus ampulliforme*, fig. 17.6*B*); and pollen organs (*Whittleseya elegans*, fig. 17.6*C*). A few whole branches have also been found. Figure 17.5 shows a beautiful example of *Neuralethopteris biformis*. The preserved frond shows thirty-one pinnules, most including some of the carbonized remains of the original plant material. Fine lines covering the pollen organs (fig. 17.6*C*) are narrow tubes that contained the pollen. All these seed-fern parts are thought to be associated with one tree-sized plant called *Medullosa* (Dilcher and Lott 2005; Dilcher, Lott, and Axsmith 2005). *Neuralethopteris pocahontas* was a related species of frond (fig. 17.6*F*). Figure 17.6*J* shows fossil impressions of *Lyginopteris hoeninghausi*, a sphenopterid seed fern. Figure 17.1 shows an artist's impression of one type of seed fern called *Neuropteris*.

HORSETAILS (SPHENOPHYTES)

Calamites is a common fossil horsetail (so called because its top resembles a horse's tail; see fig. 5.4, right side, for an artist's rendition of part of one; also figs. 5.5 and 17.1). Modern horsetails are tiny plants found near rivers and lakes; *Calamites* was tree sized. Often found in flattened form, *Calamites* is best known from casts, as in the specimens shown in figure 17.6*H* (*Calamites undulatus*) and 17.6*I* (*Calamites goepperti*). The casts form because the inside of the trunk was hollow; the casts are solidified mud or sand impressions of the inside of the trunks. The cylindrical bands are called nodes and the parallel ridges and grooves ran along the trunk. The large oval structures on the left side of figure 17.6*I* are branch scars, where the branches came out of the trunk.

Several branches of the calamitean tree are shown in figure 17.7 (left). These are assigned to *Asterophyllites charaeformis* and would have been side branches coming off the main trunk. The reproductive organ of the tree is assigned to *Calamostachys*. An example is shown in figure 17.7 (right). Like the scale trees, the calamitean tree reproduced via spores.

This completes our look at Minkin Site fossils. Now let us look to the future and elsewhere in the Black Warrior Basin.

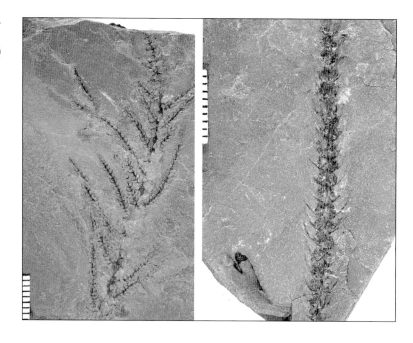

FIGURE 17.7 *Left, Asterophyllites charaeformis,* fossils of the lateral (side) branches of *Calamites;* right, *Calamostachys* sp., a cone-like reproductive organ of the *Calamites* plant.

18

The Next Steps

Having placed the Minkin Site into the context of its geology and natural history, and having described the fossils found there and how the site came to be protected by the state of Alabama, we now have to ask: Where do we go next? What can we ultimately expect from future studies? There are several issues to consider: (1) turnover and ongoing collection; (2) use of the existing database; (3) possible new excavation of the highwall; and (4) potential for finds elsewhere in Alabama. We discuss the first three in this chapter and consider the fourth topic in the next chapter.

TURNOVER AND COLLECTING

More than a decade after the dedication ceremony, amateurs are still doing the bulk of the work in collecting new specimens. High-quality trackways are still being found, because the State Lands Division sends in a bulldozer every six months to turn over the rocks and expose new material. The State Lands Division also sends out inspectors unannounced to check the site and maintain it. Because this is done systematically, the Minkin Site often looks like tilled soil. The site continues to fulfill its educational potential as a destination for schools, other interested groups (fig.18.1), and various well-known visitors such as Alabama author Homer Hickam (P. Atkinson, pers. comm. 2009). Four more track meets have been held since 2001 and more are scheduled as newly found specimens accumulate. Amateur collectors are encouraged to donate exceptional specimens to the McWane Science Center, which now maintains the largest collection of Minkin Site fossils in the state.

FIGURE 18.1 A group of visitors enjoys collecting fossils from the state-protected Union Chapel Mine at the Minkin Paleozoic Footprint Site.

The idea of keeping the rocks at a former strip mine "stirred up" is unusual, because it is the antithesis of mine reclamation and is being done strictly for the benefit of collectors and researchers. This is a great service to science and education and allows the site to continue to be productive. The site would have yielded few new specimens and would have been covered with vegetation long ago if not for this periodic reshuffling, which also prevents rocks from sitting exposed long enough to disintegrate. It may eventually be necessary to dig deeper in future turnovers in order to access rocks not reached by earlier cuts. If this is done, the potential for rare and interesting finds should be greatly enhanced.

USING THE FOSSIL DATABASE

The issue of how to use the fossil database belongs in the domain of researchers. Having so much material in institutional settings allows researchers from anywhere in the world to access the specimens directly. But even if potential researchers cannot visit Alabama, they can use the online Photographic Trackway Database to view much of the material. Online documentation also allows interested researchers and laypeople to interact with the database to examine and classify the material. A new trackway database, prepared in 2012, allows people to select interpretations of the ichnospecies, substrate characteristics, and other details at the click of a button.[1] Interactive and interpretive web design allows the convenient preliminary examination of a large database that would otherwise be very time consuming to examine directly in a museum. This approach facilitates follow-up research, including statistical studies. An interactive database can be easily updated with the latest interpretations. For example, like the Union Chapel Monograph, the original online database misinterpreted monuran insect walking and jumping traces and likely misrepresented terrestrial scorpion traces as the xiphosuran trace fossil *Kouphichnium*.

Ultimately, doing research on the trace fossils means examining actual specimens rather than just photographs. Photographs can suffer optical distortions, and it can be difficult to tell in a photograph whether the tracks in a specimen are in negative epirelief or positive hyporelief. The most reliable measures of trackway properties (pace, stride, straddle, pace angulation, manus/pes digit lengths, etc.) are made from the rocks themselves. Professional paleontologists involved in the earlier phases of study of the site recognize there is still much more to do. Spencer Lucas notes that the invertebrate trace fossils need more detailed study, and that scrutiny of *C. cobbi* trackways is needed to better understand how the animal interacted with the substrate as it moved in different ways (S. Lucas, pers. comm. 2010). Pioneering studies of *Undichna* from other parts of the world make it clear that we have only scratched the surface with the Minkin Site *Undichna* material. Well over a hundred specimens have been collected and only a few have received more than a cursory examination. New insights into the kinds of fish responsible for the traces, their behavior, and their interaction with the environment are possible with further study (Anderson 1976; Martin and Pyenson 2005; Seilacher 2007; Martin, Vazquez-Prokopec, and Page 2010). Another promising research area concerns the relationships among the organisms that left their traces at the Minkin Site. We have examples of trackways that cross, trackways that turn away from or go around obstacles, and organisms that appear to travel together. Two Union Chapel Monograph papers have been published about the plant fossils found at the Minkin Site (Dilcher and Lott 2005; Dilcher, Lott, and Axsmith 2005), and there is potential for paleoecological studies, including a search for insect bite marks on plants, which have, at other sites, provided striking evidence of insect-plant interactions.

DISSECTING THE HIGHWALL

While periodic turnover and deeper cuts will help yield new specimens, the stratigraphic context of these specimens will still be unknown. However, we can be certain that stratigraphic context is available from the highwall. Thus, many involved with the project suggest systematically dissecting it. This is not as far-fetched as it may seem. As discussed in chapter 9, Jack Pashin carried out a detailed stratigraphic analysis of the highwall and concluded that the bulk of UCM tracks came from layers 3.5 to 5.5 m (12 to 18 feet) below the exposed coal seam (called the New Castle in fig. 9.2B). This interval lies some 25 m (82 ft) below the top of the highwall. Thus, to reach it, a massive amount of overburden would have to be removed.

From a scientific point of view, the trackways that should be in that wall are more important than the ones found in the spoil piles, because the latter have lost the context in which they formed. We cannot determine precisely where a rock in a pile came from, and therefore how it connects to other fossils and sedimentary features like ripples or gas bubbles. But if we could excavate the highwall, the situation would be different. Whole bedding planes are preserved there in situ, and if we could expose the main track-bearing layers, we would have complete context and relationships preserved. Has something like this ever been done at a coal mine simply for the sake of science? As far as we know, the answer is no, at least in the United States. Interviewees for this book think an excavation of the highwall is feasible but may be expensive unless time and equipment are donated. Any such excavation would need to be very closely supervised by someone with a strong and genuine interest in the science (J. Lacefield, pers. comm. 2009). Dolores Reid considered it only a matter of convincing someone to do it, hoping that "all the mining companies in that area . . . would pitch in and do the drilling and blasting for you" (D. Reid, pers. comm. 2010).

The most effective approach would probably be to terrace the highwall, gradually exposing layers with increasing depth. Terracing would facilitate construction of a building or warehouse over the track site, as was done at Dinosaur National Monument in Utah. To keep this possibility alive as long as possible, Prescott Atkinson believes that Representative Aderholt's Union Chapel Mine bill should be reintroduced to both houses of Congress. This is because the current situation defers reclamation of the mine area for only forty years, not indefinitely. "If [the site] could sit for 200 years, it would still be valuable because we know that underneath the highwall there are still high quality trackways and interesting sediments and relationships we could study," he said (P. Atkinson, pers. comm. 2009).

Ashley Allen expressed similar sentiments. He can see excavating the highwall layer by layer and extracting segments that can be reconstructed in a repository, such as a large on-site warehouse or museum. This would be like what Jerry MacDonald did, except no one would have to carry slabs out on their backs! Allen pointed out that highwall excavation would not only give more complete information on specific tracks and behaviors, but would also provide information on just how many track-bearing layers are actually present and how they differ from one another, questions difficult to answer with only spoil-pile slabs. "I think if we can go through layer by layer, we stand to learn so much more out there," he said (A. Allen, pers. comm. 2010).

Jim Griggs pointed out that the State Lands Division owns only a limited amount of land on the other side of the highwall, and the total amount of Minkin Site land owned by the Division is only four acres. He feels the task would be very expensive, and he also noted that when overburden is removed, you have to place it within those four acres. Thus, the best approach in his view is to exhaust the already exposed areas, and then some time down the line excavate the highwall and place the overburden on the exhausted ground. The Minkin Site would not be a "mine" but a "piece of property we are digging into," he said (J. Griggs, pers. comm. 2010).

Spencer Lucas's view is that the highwall "should be reopened in a big way," but only if we "had a plan for where all the fossils will go." He recommends that if there is no public institution in Alabama that can house the bulk of the collection, then arrangements should be made with a consortium of institutions nationwide that would be willing to take some fraction of it. The critical thing is to make sure that all the fossils are publicly available and preserved indefinitely. He believes that only a controlled collection, such as would be obtained by excavating the highwall, can precisely establish associations between different traces and surfaces. He also noted that while doing such an excavation strictly for science might be unprecedented in the United States, it has been done in France and Germany. Although removing 25 m (82 ft) worth of overburden sounds difficult, he said, "it's nothing for a big machine. It's just money and a machine, that's all. . . . It would require somebody to bring it all together, because you need somebody who knew how to mine, who had a totally technical background in that area, the funding to underwrite it, and then some group of people who had the scientific eye for what the product would be. So you would have to bring those 3 things together. It would require a bit of orchestration" (S. Lucas, pers. comm. 2010).

There can be no doubt that excavation of the highwall at the Minkin Site would be an exciting venture. It would open a window to the coal age that would be far clearer than rock-pile yields could ever allow. It would also be one of the most ambitious and important paleontological projects ever attempted in Alabama.

19

Megatracksite

Trace fossils from the Minkin Paleozoic Footprint Site do not exist in isolation. Tetrapod trackways have been documented in other locations in Walker County, and all appear to be associated with the Mary Lee coal zone. Most of the known tracksites are near Carbon Hill, including the Galloway No. 11 Mine. It is important to ask what the broader implications of these multiple sites might be. Where do they all lie in the geological time sequence and how do they relate to paleoenvironments of coal-age rocks in Alabama? We can begin to answer these questions because of a new tracksite near Carbon Hill.

The Crescent Valley Mine (CVM) is the most recently discovered prolific tracksite in Alabama. It is located 37 km (23 mi) west of the Minkin Site, in an area a little southwest of Carbon Hill. Like the Minkin Site, the CVM began as a strip mine, with large piles of rock, pits, and exposed highwalls. Comparing the rock layers at the CVM with those exposed at the Minkin Site reveals an exciting possibility: Walker County may contain a *megatracksite*. This is a set of tracksites distributed across an extensive region in which all the tracks come from the *same stratigraphic interval* (meaning they are all about the same age).[1] If a megatracksite does exist in Walker County, each known site would be sampling a different point in space within a single interval of time, allowing us to map the ancient environments that once defined a common ecosystem.

In this chapter, we describe how the CVM became a part of our story, what kinds of fossils the mine has produced compared to the No. 11 and Minkin sites, and how it led to the concept of a megatracksite in Walker County.

VISIT TO CARBON HILL

From 2008 to 2012, the CVM was an active mine in an area just off the northern edge of the old Galloway No. 11 Mine (figure 7.2). A prime target of this new operation was the Jagger coal bed, the same bed whose roof shale had yielded the discovery tracks of Blair, Miller, Aldrich, and Jones in the 1920s (chapter 7). Because the CVM was sampling the same layers of rock as the old mine, the new operation opened up the possibility of a direct connection to those historical discoveries. More importantly, an active surface mine allows a view of the rock layers that an underground mining operation generally cannot provide. Surface mines expose highwalls that allow a broad view of the stratigraphic context and geological history of a site, and help us correlate different sites layer by layer.

Trackways at the CVM were brought to our attention in March 2011, when coauthor Ron Buta decided to visit Carbon Hill to view the site of the old No. 11 Mine, mainly to see what the area was like today and to gain some historical perspective. Richard Carroll, a coal geologist and paleobotanist at the Geological Survey of Alabama, had access to a 1935 map of the No. 11 Mine and also wanted to see the site. This map had accurate information on the longitude and latitude of the former mine entrance and was used to make the schematic shown in figure 7.2. Buta had never been to an abandoned underground mine before. He thought that after eighty years, the entrance to the mine would still be visible and boarded up, with railroad tracks coming out, and that it would be necessary to get permission from the current landowner to explore the site. The reality was quite different. Using Global Positioning System (GPS) technology, the location of the old entrance was pinpointed among mysterious concrete ruins in a forest just off a nearby road. Only scattered bits of coal and shale betrayed the activity that was once the livelihood of the people living in the area. The entrance was completely gone, and there were no buildings or houses nearby.

It was during this visit that the CVM was noticed just north of the No. 11 Mine ruins. Large piles of rock and soil were visible from the road and the sound of mining equipment could be heard. The mine apparently got its name from its location on W. Frank Cobb Sr.'s old Crescent Valley Farm (W. Frank Cobb III, pers. comm. 2011). Recall that the trace fossil *Cincosaurus cobbi* was named after Cobb for his generosity in allowing excavations of tracks from the No. 11 Mine. Walter Jones had mentioned in Museum Paper 9 that *C. cobbi* traces could be found almost anywhere in the No. 11 Mine tunnels. Could there be tracks in those freshly exposed rock piles across the road? Buta wanted to know, and the best way to find out was to ask the miners directly.

Leland Lowery is a veteran coal miner who was the foreman of the CVM at the time of Buta's visit. On seeing the unexpected visitors at the mine gate, he drove over to ask why we were there. Buta told him about the fossil tracks book project and explained that they were there to view the site of the old No. 11 Mine where the ancient tracks were first discovered in Alabama. Buta then showed him a copy of the 1930 Aldrich and Jones paper and asked whether the miners had seen such fossils during their operations. Lowery said he had indeed seen such things but he did not know much about them. He directed the visitors to another miner, Donny Williams, who was an avid collector of coal-age fossils, including trackways.

Williams is a nearly forty-year mining veteran who operated the Kansas No. 2 Mine, located about 8 km (5 mi) from the CVM (site KM2; noted below). Williams's mine yielded vertebrate tracks whenever the Jagger coal bed was exposed. At such times, he would have his miners place the overburden in a special area where he could later look for these fossils. Williams told us that he was at the CVM when mining operations started there in 2008, and in an intriguing repeat of history, he recounted how he had seen trackways in the ceiling of an exposed coal-excavated cavity, a relic of an old underground mining operation.[2] He recalled that the footprints in the cavity were not small like typical *C. cobbi* tracks but were much bigger, about the size of a human hand. He said the cavity was very close to the surface, and that the Jagger seam was like a visible outcrop that could be followed farther into a hill.

The photographs in figure 19.1*A–C* show what is likely to be the actual cavity Williams saw. It was on the north side of the mine and was likely not originally part of the No. 11 Galloway mining operation.[3] Large tracks (which appear to be of the ichnotaxon *Attenosaurus subulensis*) were seen on a collapsed slab in the cavity (marked *X* in fig. 19.1*A*) and are shown close up in figure 19.1*B*. The circumflexes in

FIGURE 19.1 (*A*) A photograph taken in 2008 of an old coal-excavated cavity that was found close to where mining operations began at the Crescent Valley Mine. The *X* marks a slab inside the cavity that had collapsed from the ceiling and that showed large tetrapod footprints. The arrow points to the top of the Jagger coal bed. (*B*) A close-up view of the collapsed slab, showing at least five large tracks of an animal moving toward the left. The tracks are of the ichnospecies *Attenosaurus subulensis*. In addition to the five tracks, two deep undertracks (showing only two toes each) of a second animal are seen moving to the right. These tracks may have been made several days later. The yellow color on parts of some of the rocks at the mine (such as this slab) is caused by natural staining with iron minerals. (*C*) The ceiling above the slab shows counterimpressions of the slab tracks, in positive hyporelief. The two most obvious ones are indicated by the short arrows.

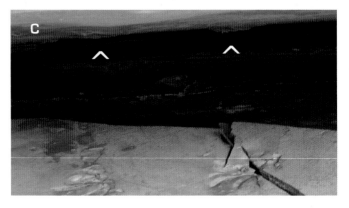

figure 19.1*C* point to track counterimpressions in the ceiling above the collapsed slab. Ceiling tracks are always protrusions. They are the sediment that filled tracks, not the tracks themselves. Williams noted that as the mining operation continued, more old cavities like this one were exposed, and that the highwall at the CVM occasionally had a honeycombed look when several were exposed at the same time.

After meeting with Williams, Buta returned to the CVM the following week, and Lowery took him on a grand tour. The first spot he showed was where Jagger coal was being freshly mined. A large area had been cleared of overburden, and a mining ripper was installed on top of the exposed coal seam, getting the coal out. Several large cavities in the highwall surrounding this area indeed gave the mine a honeycombed look (fig. 19.2). None of these cavities was deep, and no tracks were visible in the ceilings.[4] A primary goal of the Crescent Valley mining operation was to extract the coal left in the pillars separating the old cavities. Underground mining gets only 70 percent of the coal, so the new surface mine was busy retrieving the remaining, economically significant 30 percent. The arrow in figure 19.1*A* points to the top of a pillar of Jagger coal.

At a spot with an exceptional view of the highwall farther into the mine (fig. 2.2*A*), one could see the three beds of coal being mined: the

FIGURE 19.2 A pit at the Crescent Valley Mine showing the remains of several large, coal-excavated cavities from past underground mining. The reason the cavities are not at the same level is that the Jagger coal bed has faults (discontinuities) in that particular area. The cavities give the highwall a honeycombed look. Photo taken in April 2011.

Mary Lee, Blue Creek, and Jagger. The two large caves at the bottom of the photograph are more old cavities excavated in the Jagger bed. Three weeks after the picture in figure 2.2*A* was taken, the wall was not recognizable and the cavities were less conspicuous. In strip mines, the scenery changes quickly!

The CVM highlights well how coal mining has changed over eighty years. The No. 11 underground mine focused mainly on Jagger coal, which miners followed wherever it went. Also, although the No. 11 coal excavations were generally less than a hundred feet below the surface, the equipment needed to remove the overburden did not exist in the 1920s. At that time, strip-mining was a new method still under development. Today, former underground coal mines are converted into productive strip mines. The strip-mining allowed the excavation of not only leftover Jagger coal, but also shallower Mary Lee and Blue Creek coal.

Naturally, after seeing all this, Ron wondered what had become of the ceiling tracks that Donny Williams had mentioned. When the new mining operation started, the rooms that had been cleared of Jagger coal long ago were still visible, and in some, tracks were observed in the ceiling. But now these rooms, as well as the pillars that defined them, are mostly gone, consumed by the surface mining operation.[5] We can

FIGURE 19.3 CVM 634: At least two *C. cobbi* trackways cross this large slab. The best-defined footprints are of the manus. In these undertracks, the pes prints show only two digits.

nevertheless determine that the ceiling tracks described in 1930 and those noticed eighty years later were more than a mile apart.

With Lowery's permission, Buta visited the CVM periodically over the next eighteen months, and Jack Pashin measured the section, as he had at the Minkin Site. The site proved very rich in trace fossils, especially vertebrate trackways, and yielded exceptional specimens. Not surprisingly, the CVM ichnofauna is very similar to that from the No. 11 Mine. Figures 19.3–19.9 show a sampling of what was found, including some invertebrate traces that were rare at the Minkin Site.

Most of the small vertebrate trace fossils collected are *C. cobbi*. The specimen in figure 19.3[6] has two sets of overlapping tracks. It is possible that two similar animals were moving together or that the two sets of tracks were made by a single individual. As is typical of *C. cobbi* traces in general, the best-defined tracks are of the manus, where at least four toe prints are seen. In these undertracks, the pes prints show only two digits.

In another *C. cobbi* specimen (fig. 19.4), the pes prints are noticeably larger than the manus prints and show three long and relatively straight traces of digits with a row of dots at the base. Most of the manus prints show four digits. Notice that the tracks are paired, with the pes prints behind the manus prints. In figure 19.5, by contrast, the pes prints are in front. These differences have to do with gait and walking speed. The pace angulation is larger in figure 19.4 because the animal was taking longer strides, and probably walking faster and more upright.

FIGURE 19.4 CVM 348–9: *C. cobbi* trackway where the pes prints are significantly larger than the manus prints. In these undertracks, the pes prints show only three digits. The tracks are paired, with the pes behind the manus.

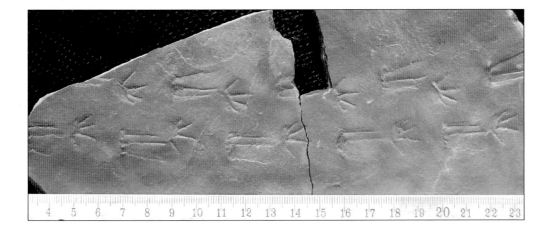

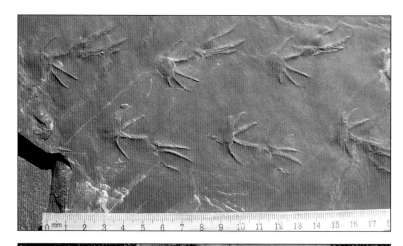

FIGURE 19.5 CVM 786: In these undertracks, the manus prints show five digits and the pes prints show three. The tracks are paired as in figure 19.4, except that in this case, the manus prints are behind the pes prints.

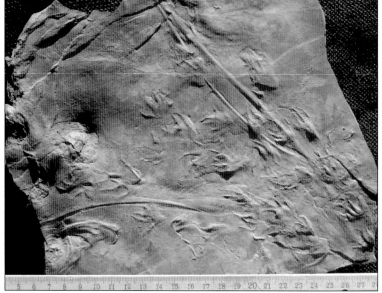

FIGURE 19.6 CVM 468: Tetrapod tracks made at the original surface, from multiple individuals. The ichnospecies is uncertain because the tracks are distorted from movement in what was likely very wet mud. Note conspicuous tail-drag marks, which are not generally found in undertracks.

Surface tracks are rare at the CVM, but some interesting examples were found. One is shown in figure 19.6. We know these were made at the original surface because of the conspicuous tail impressions, which are usually not found in undertracks. Even though surface tracks are highly prized, these do not tell us much about the animals. The tracks are stretched and blurred, and there is considerable overlap from multiple individuals. Nevertheless, figure 19.6 shows a busy scene and possibly even group behavior.

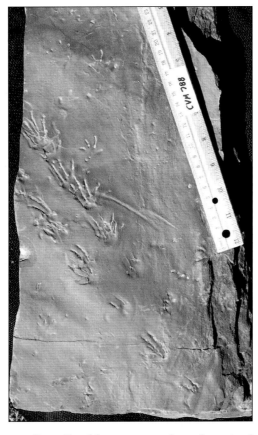

FIGURE 19.7 CVM 788: A well-defined *C. cobbi* trackway where the animal walked from drier substrate into wetter substrate. Tracks at the beginning of the trackway are ill defined and considerably shorter than tracks farther along. In the wetter area, manus and pes tracks overlap. An unrelated set of *C. cobbi* undertracks can be seen near the main part of the trackway.

Figure 19.7 shows an excellent *C. cobbi* specimen where the animal clearly walked from drier to wetter ground. The tracks begin ill defined with shortened digits and become progressively more conspicuous and complicated because manus and pes overlap. The animal was likely walking into shallow water. Interestingly, another set of *C. cobbi* undertracks cross the main set. These must have been made later, after the other set was covered.

Large tracks identified as *Attenosaurus subulensis* are the only other kind of vertebrate tracks found at the CVM. In figure 19.8, the largest track is a pes showing a clear "thumb" and three long digits, with possibly a trace of the fifth digit; it is roughly the size of a human hand. The smaller track is a manus showing only three clear digits. Both are undertracks. *A. subulensis* is rare, wherever it is found, compared to the tracks of small tetrapods.

FIGURE 19.8 CVM 641: *Attenosaurus subulensis.* The largest track is a pes print, while the smaller is a manus print. Part of a second pes print can also be seen. All are under-tracks.

The most common invertebrate trace found at the CVM is called a chevronate trail because it consists of a pattern of nested *V*s, which is called a chevron (fig. 19.9). This kind of trail is associated with two similar forms at the CVM,[7] and all three are absent or extremely rare at the Minkin Site. The most common invertebrate ichnogenera in the *Cincosaurus* beds at the Minkin Site, *Treptichnus* and *Arenicolites*, are also fairly common at the CVM.

There are a number of important absences at the CVM. For example, Buta found no definite amphibian trackways after nearly thirty visits to the site. This is consistent with Aldrich and Jones, because no amphibian tracks are illustrated in their article. In contrast, the small temnospondyl amphibian trace *Nanopus reidiae* is the most common vertebrate trackway at the Minkin Site. Also not found at the CVM were any *Undichna* (fish trails) or *Kouphichnium* (horseshoe crab) tracks. *Stiaria* and *Tonganoxichnus* (traces made by monuran insects) and *Diplichnites* (millipede trackways) were found but were very much rarer than at the Minkin Site. While it is possible that more in-depth study of the CVM might expose amphibian, *Undichna*, or *Kouphichnium* traces eventually, it is also possible that the water at the CVM might have been

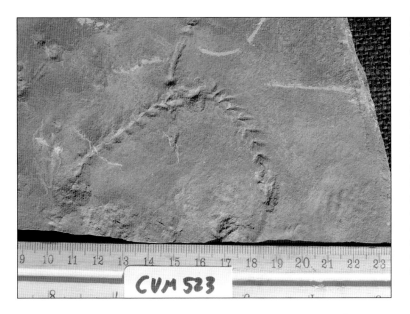

FIGURE 19.9 CVM 523: A very common type of invertebrate trace characterized as "chevronate" because of the succession of *V*-shaped patterns. Although not yet assigned to an ichnospecies, these traces have been attributed to the activities of juvenile horseshoe crabs by Buta at al. (2013). They also resemble the ichnogenus *Protovirgularia*, thought to have been produced by a bivalve (an animal with two hinged parts).

salty enough to exclude amphibians, which prefer fresh or only slightly brackish water (Buta et al. 2013). The scarcity of terrestrial invertebrate traces of millipedes and monurans supports this interpretation. The chevronate trails may have been made by juvenile horseshoe crabs (Buta et al. 2013). Overall, more than a thousand slabs bearing trace fossils were collected and documented from the site.[8]

MAPPING THE MEGATRACKSITE

Lockley and Meyer (2000) define megatracksites as regionally extensive track-bearing layers or surfaces as much as hundreds or thousands of square kilometers in area. In dinosaur track studies, a megatracksite is often characterized as a "dinosaur freeway." The key to a megatracksite is finding widely distributed tracksites from the same stratigraphic interval (that is, the same layers in the rock sequence, implying the same age). A megatracksite has considerable scientific value. As we have noted, it can provide a broader picture of the paleoecology of an area and allow us to map extensive ancient shorelines. Exploring a megatracksite requires a considerable investment of time and effort, and attention to details that connect individual sites.

The locations of tracksites in Walker County are shown in figure 19.10. The Minkin Site anchors the far eastern end of an almost linear distribution of sites where the Westphalian-age tetrapods left their marks. These tracksites cover a fairly large area, about 400 km² (150 mi²). However, the map in figure 19.10 is not likely to be complete because it shows only the thirteen tracksites presently known.[9] The APS discovered tracksites at the Fern Springs Mine (FSM) and Sugar Town Mine (STM) in 2005–2010. The FSM was a discontinued mine that had lain unreclaimed for many years, and scattered rocks were badly weathered. The STM was still active at the time of the APS visits and yielded an extensive collection of mostly plant fossils with a few tracks. APS members donated these to the McWane Science Center. Much of this material has not been studied or documented. Other mines, like the Hope Pit Mine (HPM), the Kansas Mine (KM), and the Gateway Malls–Hope Galloway Mine (GMI-HPM) (Liu and Gastaldo 1992; Gastaldo et al. 1990; J. Pashin, pers. comm. 2013), were found to be rich in trackways many years before the UCM discovery, but little documentation was made for these. The Holly Grove Mine (HGM) yielded several specimens catalogued at the Alabama Museum of Natural History. Even though the Cedrum Mine (CM) was large, it yielded few tetrapod trackways (chapter 1). Cedrum was much richer in invertebrate traces (Lacefield and Relihan 2005). All these sites, except the Minkin Site, are badly deteriorated and have been reclaimed, or soon will be.

There were likely other mines (and even construction sites) in Walker County that yielded tracks over the years, but we do not know anything about them. No systematic record is available, especially for mines or construction sites that were active before 1984, the year the BPS was founded. Most of the identified tracksites lie in western Walker County in the Carbon Hill and Kansas area. A few tracksites are scattered across Walker County to the east, with the Minkin Site being the easternmost. Recall from chapter 3 that tracksites point to the previous existence of a shoreline in an area. The known sites in Walker County follow a relatively narrow pattern oriented WNW-ESE that suggests a continuous swampy shoreline at least 45 km (28 mi) long.

If Walker County does in fact host a real megatracksite, then individual sites will likely have similar ichnofaunas, and stratigraphic analysis

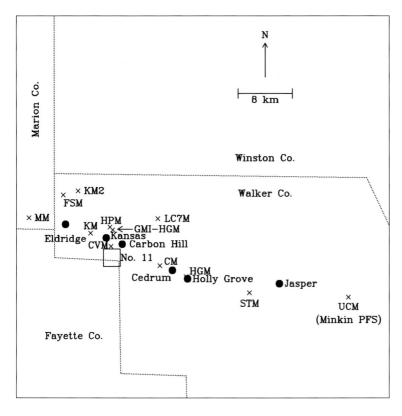

FIGURE 19.10 Known tracksites in Walker County, Alabama, relative to the cities of Eldridge, Kansas, Carbon Hill, Holly Grove, and Jasper. The mines shown are: CM = Cedrum Mine; CVM = Crescent Valley Mine (active 2008–2012); FSM = Fern Springs Road Mine; GMI-HGM = Gateway Malls–Hope Galloway Mine; HGM = Holly Grove Mine; HPM = Hope Pit Mine; KM = Kansas Mine (J. Lacefield's 1990s site); KM2 = Kansas Mine No. 2 (active in 2011); LC7M = Lost Creek No. 7 Mine; MM = L. Massey Mine; No. 11 Mine = Galloway Coal Company mine where first tracks were documented; square indicates rough actual size; STM = Sugar Town Mine (active in 2010); UCM = Union Chapel Mine (Steven C. Minkin Paleozoic Footprint Site)

of the exposed highwalls will show that the tracks come from the same rock layers. The Minkin Site and the CVM are the only tracksites for which both detailed ichnofaunal and stratigraphic information has been obtained. For most of the sites in figure 19.10, little comparable information is available. The ichnofaunas of the Minkin Site and the CVM have their differences, but they are quite similar, too. Both sites yielded *C. cobbi*, *Attenosaurus subulensis*, *Treptichnus apsorum*, *Arenicolites* isp., *Stiaria* isp., *Tonganoxichnus robledoensis*, and *Diplichnites* isp.

As we described in chapter 9, the highwall at the Minkin Site currently shows only one coal seam, while that at the CVM shows at least three. The tracks at the CVM and No. 11 Mine come from strata not far above the Jagger coal, but those at the Minkin Site come from just above the Mary Lee bed (which is now below ground level; fig. 9.2*B*). Technically, this difference could imply that CVM traces are on the order of 100,000 to 200,000 years older than Minkin Site traces (because the Jagger would be two coal seams below the Mary Lee) (Pashin 2005). However, using stratigraphic detective work, and following individual rock layers across the county in detailed records from gas wells, Jack Pashin (Buta et al. 2013) concluded that what the miners called the Mary Lee coal bed at the Minkin Site was in fact *the same bed* that was called the Jagger at the CVM. If correct, this inference means the *Cincosaurus* beds are essentially the same age on opposite sides of the county, and perhaps they are the same everywhere in Walker County where vertebrate tracks are found. This important correlation, unrecognized until eighty years after tracks were first reported, opens the door to a much better understanding of life in Alabama during the coal age.

In summary, the CVM has brought to paleontological studies in Alabama: (1) a significant number of new specimens from the same area where tracks were first discovered in the state; (2) a large database of fully documented specimens that provide much-needed perspective on the Minkin Site trace-fossil assemblage; and (3) the realization that Walker County shows evidence of a megatracksite. The future of trackway studies in Alabama will involve the exploration of the megatracksite as new sites (coal mines, building sites, and road cuts) are opened up to investigation.

Epilogue

It might at first be hard to believe that Alabama was once located south of the equator, that it was partly covered by tropical swamps with gigantic insects and bizarre scaly trees, that it harbored an exceptionally high mountain range, and that it was home to some of the first reptiles ever to evolve. However, geologists have excellent evidence to suggest that this was so. During the "coal age" 313 million years ago, the land, the water, the sky, and the plant life were very different from what they are today. The trace fossils of Walker County open our minds to a vanished world and tell us about ancient animals that would be completely unknown if all we could depend on was fossil bones.

It is equally hard to grasp what a long period of time 313 million years is. If every one of those years became a second, they still would add up to almost ten years. Our sun has moved more than one orbit around the center of our galaxy since the coal-age animals walked the Walker County mudflats. When the coal-age world came into being, dinosaurs, the most popular and fascinating land animals that ever lived, were yet to evolve, far in the future. Paleozoic animal life rarely captures the public's imagination like dinosaurs do, but these ancient creatures were the ancestors of dinosaurs, and eventually of humans.

The successful documentation and protection of the Union Chapel Mine was an extraordinary experience for Alabama's amateur fossil collectors. How many similar groups can claim such success? The protection of the UCM for long-term study was not an accident. Luck was involved, but without the dedication and hard work of the APS, the mine would now be rock-free, rolling grassy hills. Jim Lacefield precipitated the first chapters in this saga when he introduced the BPS to fossil animal tracks in Alabama coal mines, but he was only the first of many who eventually worked together to preserve this important fossil site. Everyone involved should be proud of this accomplishment.

There can be no doubt that the UCM experience had an impact on everyone involved. More than ten years after the site was discovered, interest in fossil trackways and traces from the mine is as high as ever. Hundreds of people, including students and educators, have visited the site. Fossil trackways inspire the imagination because they are so much about life. It is easy to imagine an animal scurrying across a wet mudflat that is now laminated gray rock. We wonder where the animal was going and what it was doing. Sometimes, we can tell that it was avoiding an obstacle, or walking from land into water. Bones can never tell us such things. Trace fossils bring the world of coal-age Alabama back to life.

Coal itself and the way it is mined are exceptionally interesting aspects of the natural and commercial history of Alabama. Coal is not merely an energy source; it is the highly compressed and altered remains of magnificent forests that covered the land for tens of thousands of years. It is difficult to look at coal seams and not be fascinated, whether they are exposed at the surface or thousands of feet underground.

The success of the APS in getting the Minkin Site protected less than six years after it was discovered stands in contrast to Jerry MacDonald's experience in New Mexico, and that of others in other regions. Mac-Donald pushed hard and persistently to get track-bearing areas, and indeed a large swath of the Robledo Mountains, protected as a national monument, but the task took nearly twenty years. Meanwhile, quarrying operations continued unchecked in the Robledos, destroying much significant fossil material. When asked why it took so much longer than the Alabama effort, MacDonald replied:

> Well, the obvious thing was, boy, you had a lot of enthusiasm that helped. And I didn't. I mean, it was remarkable your turnout and your sustainability with people that were genuinely excited all the way down the line. You have to have that. First of all, you obviously have to have the enthusiasm to discover, and you had that with [Steve] Minkin. . . . He was well received, and he got immediate support from academics and other interested people that

recognized the significance of that area. . . . I didn't have any of that. I had no in-state network. . . . Nobody was interested. (J. MacDonald, pers. comm. 2009)

An important issue with Jerry MacDonald's experience is that he made his big discovery when the center of paleontology in New Mexico was being transferred from New Mexico Tech to the New Mexico Museum of Natural History and Science, a process that he said created some tension among academics in his state. In Alabama, the APS had the support of the mining company from the start. Professional paleontologists in Alabama supported the UCM rescue project by talking to state and federal officials, going into the field with BPS and APS members, providing storage facilities for UCM specimens, helping with track meets and the Permo-Carboniferous Workshop, and more. The "series of fortunate events" that happened at the Minkin Site did not unfold so fortuitously in New Mexico or, for that matter, with the Kansas Mine.

The Crescent Valley Mine experience described in the previous chapter suggests that tracksites as rich as or richer than the UCM can still be found in Walker County. Even so, it is not practical to preserve every coal mine that might have important trace fossils. Some mining companies will no doubt be happy to allow people to explore their mines for such fossils while others will not, the latter most likely because of liability issues.

There is much still to learn about the mysterious world of coal-age Alabama. We hope this book has opened a door to that world and to a greater appreciation of its relevance to life today.

Appendix 1. Timeline of Events

~313,000,000 bp (before present) Deposition of the "*Cincosaurus* beds."

1912 Galloway Coal Company No. 11 underground mine begins operations near Carbon Hill.

1927. Walter B. Jones is appointed state geologist of Alabama.

Early December 1929 Geologist Arthur Blair and Tennessee Coal, Iron, and Railroad Company man Ivan Miller visit the Southwest Slope of No. 11 Mine to view the exceptional display of ceiling tracks, at the invitation of mine general manager W. F. Cobb Sr. and chief engineer A. P. McIntosh. A portion of the display, which includes the tracks of a five-toed (manus and pes) tetrapod, is extracted and taken to Birmingham, where Walter B. Jones was on hand to see them.

Late January 1930 George Gaylord Simpson, associate curator of the American Museum of Natural History in New York City, visits Carbon Hill to view track displays and collect specimens. The tracks he collects in Carbon Hill, part of a tour of southern sites, are characterized in press releases as the most important and significant finds of his trip.

April 1930 Walter Jones submits manuscript titled *Footprints from the Coal Measures of Alabama*, with coauthor Truman H. Aldrich Sr., for publication that same year as Alabama Museum of Natural History Paper 9. The manuscript was prepared quickly in order to "keep the discovery in Alabama."

April 1932 Truman Aldrich dies at age eighty-four.

ca 1934 Operations end at the Galloway No. 11 Mine.

1961 Walter Jones retires as state geologist.

1977 Walter Jones dies at age eighty-two.

1984 Birmingham Paleontological Society (BPS) is founded by staff members of the former Red Mountain Museum.

November 1999 Seventh-grade Oneonta High School student Jessie Burton tells his science teacher, Ashley Allen, that his grandmother, Dolores Reid, owns active surface coal mines.

November 1999 Ashley Allen scouts the Union Chapel Mine (UCM), the most fossil-rich of the mines owned by Dolores Reid. Allen finds not only the usual plant fossils but also Paleozoic tetrapod trackways, fulfilling a long-term goal of his collecting efforts.

December 6, 1999 At the regular BPS meeting this month, Ashley Allen shows the remarkable Paleozoic footprints he collected from the UCM the month before and excites the group into scheduling a field trip to the site.

January 23, 2000 BPS visits the UCM for the first time (eleven people attend, including coauthor Ron Buta).

March 5, 2000 Authors visit the mine together.

March 8, 2000 Coauthor David Kopaska-Merkel and Andrew K. Rindsberg visit the mine to collect a representative sample of fossils for the Geological Survey of Alabama.

2000 The first publication on UCM trackways, by Steven C. Minkin, a licensed professional geologist from Anniston and also a BPS member, appears in the newsletter of the Alabama Geological Society.

August 19, 2000 Track Meet 1 is held at the Alabama Museum of Natural History, Tuscaloosa, cosponsored by the Birmingham Paleontological Society, Alabama Museum of Natural History, and Geological Survey of Alabama.

August 20, 2000 The *Birmingham News* publishes the first story about the Union Chapel Mine in an article titled "Fossil Hunters Follow Tracks to Distant Past," by Thomas Spencer.

October 14, 2000 Track Meet 2 is held by the BPS at Oneonta High School. Arranged by Ashley Allen.

December 2000 UCM partial reclamation.

May 12, 2001 Track Meet 3 and Plant Fest is held by the BPS at the Anniston Museum of Natural History. Arranged by Steve Minkin and Ashley Allen. "Plant Fest" means plant fossils are also photographed at this track meet.

June 2001 In a discussion with Prescott Atkinson, Steve Minkin proposes that a campaign be mounted to protect and preserve the Union Chapel Mine.

July 14, 2001 BPS meeting with Dennis Reid, son of Dolores Reid, owner of New Acton Coal Company, to discuss preservation of the Union Chapel Mine, at the University of Alabama at Birmingham campus.

July 27, 2001 "Great Track Layout" for Union Chapel Mine photographs is held at the Alabama Museum of Natural History, Tuscaloosa; the scope and outline of the Union Chapel Monograph, the "Blue Book," is first prepared here. Organized by Ron Buta.

September 18, 2001 The Alabama Surface Mining Commission and the BPS meet to discuss preservation of the UCM.

March 4, 2002 Buta purchases a new 80GB hard drive for astronomy research and allocates a few gigabytes to post the entire photographic trackway database (~2,400 images) online.

March 16–May 9, 2002 An exhibit of Union Chapel Mine fossils, prepared by Minkin and his brother, Bruce, is held at the Colburn Gem and Mineral Museum in Asheville, North Carolina. The exhibit is titled "Tracks through Time."

April 24–26, 2002 Jerry and Pearl MacDonald visit Alabama.

April 25, 2002 Jerry MacDonald speaks to the BPS about his Permian track discoveries in New Mexico and inspects dozens of UCM trackways brought by members. His presentation, titled "The Paleozoic Trackways Project," is held in room 116 of the Business Building on the campus of the University of Alabama at Birmingham.

April 26, 2002 The Birmingham Paleontological Society hosts Jerry and Pearl MacDonald at the Union Chapel Mine.

May 25, 2002 Steve Minkin, Prescott Atkinson, and Ron Buta meet with Bill Harris, director, Fourth Congressional District, to discuss the preservation of the mine and to arrange a site visit by Rep. Aderholt.

July 3, 2002 The BPS makes a presentation to Rep. Aderholt at the Union Chapel Mine, Walker County, Alabama.

July 12, 2002 Several BPS members host Kent Faulk (reporter, *Birmingham News*) and photographer Frank Couch at the UCM.

September 9, 2002 Jerry Lanning visits to discuss incorporation of the BPS, but the meeting is disrupted, preventing Lanning's discussion; the decision is made to form a new group from part of the BPS. The new group is shortly thereafter designated as a new nonprofit organization and named the Alabama Paleontological Society (APS).

October 7, 2002 The APS holds its inaugural meeting at the Vestavia Hills Richard Scrushy Library; the new group includes all who were heavily involved in the Union Chapel Monograph and the preservation effort.

October 8, 2002 Minkin, Atkinson, and Buta visit Bruce Hamrick, the Walker County commissioner.

Late 2002 Minkin, Atkinson, and Buta meet with the Jasper Kiwanis Club to make a presentation about the UCM fossils.

2003 Dolores Reid sells the Union Chapel Mine to Bob Reed.

February 6–22, 2003 Prof. Hartmut Haubold, Martin Luther University, Halle-Wittenberg, Germany, visits Alabama to research the tracks.

February 10, 2003 Prof. Haubold gives a presentation to the APS titled "The Footprints from Walker Co. in Alabama: A Key for the Carboniferous Tetrapod Ichnology."

February 20, 2003 Prof. Haubold gives a lecture titled "Permo-Carboniferous Ichno-Diversity, Concepts and Reality," on the campus of the University of Alabama.

May 2–4, 2003 A workshop and field trip on Permo-Carboniferous ichnology take place.

June 19, 2003 Atkinson makes a presentation on the urgency of preserving the Union Chapel Mine at monthly meeting of the Alabama Surface Mining Commission.

June 19, 2003 Representative Robert Aderholt (R-AL District 4) submits House Resolution 2520, titled the Union Chapel Fossil Footprint Site Preservation Act. No formal vote is taken.

June 24, 2003 Atkinson, Buta, and Billy Orick, permits manager of the New Acton Coal Mining Company, host a site visit with Randy Johnson, the surface mining commissioner, Dr. Nick Tew, the new state geologist, and Jim Griggs, the director of the State Lands Division of the Alabama Department of Conservation and Natural Resources.

October 18, 2003 Track Meet 4 is held at the Buta residence, Tuscaloosa, Alabama.

? Track meet 5 is held at the McWane Science Center.

February 20, 2004 Steve Minkin suffers a fatal fall in his home in Anniston, Alabama.

July 1, 2004 The UCM is acquired by the state of Alabama (State Lands Division of the Department of Conservation and Natural Resources).

March 12, 2005 The Union Chapel Mine is dedicated as the Steven C. Minkin Paleozoic Footprint Site. The APS receives a certificate from the Alabama State Lands Division "in recognition and grateful appreciation of its commitment and dedication to the protection of Alabama's unique geological heritage through its untiring efforts in preserving the Steven C. Minkin Paleozoic Footprint Site at Union Chapel, Jasper, Alabama." The *Discovering Alabama* episode titled "Tracks Across Time" is filmed.

July 26, 2005 *Pennsylvanian Footprints in the Black Warrior Basin of Alabama*, also called the "Blue Book" or Union Chapel Monograph, is published by the APS and the Geological Survey of Alabama.

September 14, 2008 Track Meet 6 is held at the McWane Science Center.

March 2011 Ron Buta and coal geologist Richard Carroll visit Carbon Hill to view the site of the old No. 11 Mine and accidentally find the Crescent Valley Mine, an active surface mine less than half a mile from the former entrance to the old underground mine. Buta and Carroll also visit the Kansas No. 2 Mine, about five miles away from Crescent Valley Mine. Both mines turn out to be new tracksites.

April 2011–December 2012 Buta regularly visits the CVM and amasses a database of more than a thousand specimens, photographically documenting them in the manner of the Minkin Site track meets.

Summer 2012 Jack Pashin visits the CVM and carries out stratigraphic analysis of the highwall. He compares this with his similar analysis of the highwall at the Minkin Site, and with subsurface information about the intervening area. He concludes that the "*Cincosaurus* beds" at the two sites are from the same stratigraphic interval.

2005–present The Minkin Site is visited monthly by the APS and many school groups. Rock piles are turned over every six months or so to enhance collecting.

March 2, 2013 Track Meet 7 is held at the McWane Science Center.

July 26, 2013 Buta donates all Crescent Valley Mine specimens collected from April 2011 to July 2013 (CVM numbers 1–1505) to the Alabama Museum of Natural History.

Appendix 2. Locations of Illustrated Fossil Footprint Specimens in Alabama

1. ICHNOTAXON (or kind if not identified)	2. MINE ACRONYM	3. FIELD NUMBER	4. INSTITUTION	5. CATALOG NUMBER	6. FIGURE NUMBER
Invertebrate	UCM	67	ALMNH	PV2005.0007.0091	16.14B
Invertebrate	UCM	1272	MSC	N/A	16.15C
Invertebrate	UCM	1311	MSC	MSC 27873	16.14C
Invertebrate[1]	UCM	1881	PC	N/A	16.15C
Invertebrate	UCM	1488	MSC	MSC 27968	16.14C
Agaeoleptoptera uniotempla	UCM	2369	ALMNH	PI 2010.0004.0001.002	16.15A
Anniedarwinia alabamaensis[2]	UCM	1076A	ALMNH	PI2005.0002.0176.001	8.5, 16.15A
Arborichnus repetitus	UCM	743	MSC	N/A	16.11A
Arborichnus repetitus	UCM	1153	ALMNH	PV2014.0015.0004	16.11A
Arborichnus repetitus	UCM	2387	MSC	N/A	16.11B
Arborichnus repetitus	UCM	2449	MSC	N/A	16.11C
Arborichnus repetitus	UCM	2539	MSC	N/A	16.11B
Arborichnus repetitus	UCM	3780	PC	N/A	16.11C
Arenicolites longistriatus	UCM	2038	ALMNH	PI2005.0002.0010	16.10A
Bivalvia	CVM	523	ALMNH	PI2013.0013.0023	19.9
Diplichnites gouldi	UCM	666	ALMNH	PI2005.0002.0069	16.8A
Diplichnites gouldi	UCM	953	PC	N/A	8.12
Diplichnites gouldi	UCM	1267	MSC	MSC 27964	16.8B
Diplichnites gouldi[3]	UCM	N/A	MSC	N/A	16.8C
Dunbarella isp.	UCM	2742	PC	N/A	16.12B
Dunbarella isp.	UCM	637	PC	N/A	16.12B
Kouphichnium isp.	UCM	117	PC	N/A	16.6D
Kouphichnium isp.[4]	UCM	331	MSC	MSC 35084	1.2, 16.14A

1. ICHNOTAXON (or kind if not identified)	2. MINE ACRONYM	3. FIELD NUMBER	4. INSTITUTION	5. CATALOG NUMBER	6. FIGURE NUMBER
Kouphichnium isp.	UCM	388	MSC	MSC 27679	1.2
Kouphichnium isp.	UCM	437	PC	N/A	16.6A
Kouphichnium isp.	UCM	1070	ALMNH	PI2005.0002.0177	8.10A
Kouphichnium isp.[5]	UCM	1071	ALMNH	N/A	8.10B
Kouphichnium isp.	UCM	1268	ALMNH	PI2005.0002.0014	16.6C
Kouphichnium isp.	UCM	1505	ALMNH	PI2005.0002.0070	16.6B
Octopodichnus isp. or Paleohelcura isp.	UCM	1368	ALMNH	PI2005.0002.0076	16.13B
Octopodichnus isp. or Paleohelcura isp.	UCM	1377	ALMNH	PI2005.0002.0017	16.13A
Rusophycus isp.	UCM	3781	PC	N/A	16.12A
Rusophycus isp.	UCM	1548	MSC	N/A	16.12A
Stiaria isp.	UCM	485K	PC	N/A	8.12
Stiaria isp.	UCM	1740	ALMNH	PI2005.0002.0065	9.4
Stiaria isp.	UCM	1748	MSC	N/A	16.1A
Stiaria isp.	UCM	1749	ALMNH	PI2005.0002.0073	16.1C
Stiaria isp.	UCM	1758	ALMNH	PI2005.0002.0052	16.1B
Treptichnus apsorum	UCM	417	PC	N/A	16.10C
Treptichnus apsorum	UCM	448	ALMNH	PV2005.0007.0076	16.10B
Treptichnus apsorum	UCM	2026	ALMNH	PI2005.0002.0112	16.10B
Treptichnus apsorum	UCM	5031	MSC	MSC 27608	16.10C
Trigonotarbid spider	UCM	2281	ALMNH	PI2005.0002.0175	16.15B
Tonganoxichnus robledoensis	UCM	1060	PC	N/A	8.11, 16.3B
Tonganoxichnus robledoensis	UCM	1735	ALMNH	PV2005.0007.0015	16.3C
Tonganoxichnus robledoensis	UCM	2466	MSC	N/A	16.3A
Tonganoxichnus robledoensis/ Stiaria isp.	UCM	1349	MSC	MSC 27769	16.4A
Tonganoxichnus robledoensis/ Stiaria isp.	UCM	1410	ALMNH	PI2005.0002.0061	16.4B
Tonganoxichnus robledoensis/ Stiaria isp.	UCM	5000B	MSC	N/A	16.4C

1. ICHNOTAXON (or kind if not identified)	2. MINE ACRONYM	3. FIELD NUMBER	4. INSTITUTION	5. CATALOG NUMBER	6. FIGURE NUMBER
Undichna isp.	UCM	1348	ALMNH	PV2005.0007.0033	15.11C
Undichna isp.	UCM	1728	PC	N/A	15.11A
Undichna isp.	UCM	1731	PC	N/A	15.11B
Undichna isp.	UCM	N/A	MSC	N/A	15.12
Vertebrate	UCM	26	ALMNH	PI2005.0002.0180	15.14A
Vertebrate	UCM	1034	PC	N/A	15.14C
Vertebrate and Invertebrate	UCM	680	ALMNH	PV2005.0007.0047	15.14B
Attenosaurus subulensis	UCM	24	ALMNH	PV2005.0007.0050	15.9B
Attenosaurus subulensis	UCM	123	PC	N/A	15.9D
Attenosaurus subulensis	UCM	124	ALMNH	N/A	8.2
Attenosaurus subulensis	UCM	1074	PC	N/A	15.9A
Attenosaurus subulensis	UCM	4085	MSC	N/A	15.9D
Attenosaurus subulensis	N/A	N/A	ALMNH	PV1985.0001.0013	1.1
Attenosaurus subulensis	CVM	641	ALMNH	PV2013.0012.0596	19.8
Attenosaurus subulensis	CVM	209	ALMNH	PV2013.0012.0597	3.6
Attenosaurus subulensis[6]	CVM	210	ALMNH	PV2013.0012.0598	3.6
Attenosaurus subulensis	CVM	N/A	ALMNH	N/A	3.7
Attenosaurus subulensis[7]	UCM	UCM 1621	PC	N/A	8.5B
Attenosaurus subulensis	UCM	N/A	MSC	MSC 33968	15.9C
Cincosaurus cobbi	UCM	18	ALMNH	PV2005.7.253	8.4
Cincosaurus cobbi	UCM	109	MSC	MSC 36453	3.1
Cincosaurus cobbi	UCM	206	PC	N/A	15.1A
Cincosaurus cobbi	UCM	254	PC	N/A	15.1B
Cincosaurus cobbi	UCM	263	PC	N/A	3.9
Cincosaurus cobbi	UCM	278	PC	N/A	1.2
Cincosaurus cobbi	UCM	451	PC	N/A	15.1C
Cincosaurus cobbi	UCM	493	PC	N/A	15.1D
Cincosaurus cobbi	UCM	1075	ALMNH	PV2005.0007.0131	8.7
Cincosaurus cobbi	CVM	146	ALMNH	PV 2013.0012.0347	3.3A,B
Cincosaurus cobbi	CVM	147	ALMNH	PV 2013.0012.0346	3.3C,D
Cincosaurus cobbi	CVM	650	ALMNH	N/A	3.7
Cincosaurus cobbi	CVM	634	ALMNH	PV2013.0012.0314	19.3
Cincosaurus cobbi	CVM	348-9	ALMNH	PV2013.0012.0530	19.4

1. ICHNOTAXON (or kind if not identified)	2. MINE ACRONYM	3. FIELD NUMBER	4. INSTITUTION	5. CATALOG NUMBER	6. FIGURE NUMBER
Cincosaurus cobbi	CVM	786	ALMNH	PV2013.0012.0594	19.5
Cincosaurus cobbi	CVM	468	ALMNH	PV2013.0012.0504	19.6
Cincosaurus cobbi	CVM	788	ALMNH	PV2013.0012.0595	19.7
Cincosaurus cobbi	CVM	N/A	ALMNH	N/A	3.3
Cincosaurus cobbi[8]	N/A	N/A	ALMNH	PV1985.0001.0027	4.1, 7.3
Matthewichnus caudifer	UCM	285	MSC	MSC 26502	8.3
Matthewichnus caudifer	UCM	469	PC	N/A	15.5
Nanopus reidiae	UCM	249	PC	N/A	8.8
Nanopus reidiae	UCM	312	PC	N/A	15.7C
Nanopus reidiae	UCM	357	PC	N/A	15.7A
Nanopus reidiae	UCM	368	PC	N/A	15.7B
Nanopus reidiae[9]	UCM	677	ALMNH	PV2005.0007.0247	8.9
Nanopus reidiae[10]	UCM	1141	ALMNH	PV2005.0007.0124	4.1
Nanopus reidiae[11]	UCM	1142	ALMNH	PV2005.0007.0123	8.6
Nanopus reidiae	UCM	1797	ALMNH	PV2005.0007.0093	3.8
Nanopus reidiae/ Attenosaurus subulensis[12]	UCM	N/A	ALMNH	N/A	cover
Notalacerta missouriensis	N/A	N/A	ALMNH	PV2015.0001.0001	15.3
Plant fossils	UCM	N/A	UF	(RIGHT) UF 34373A; (LEFT) UF 34011	17.7
Multiple plant fossils	UCM	(A) P 158; (E) P 153; (G) P 162; (H) P 176	UF	(B) UF 3687; (C) UF 36908; (D) UF 33933; (F) UF 34025; (I) UF 3392; (J) UF 34039	17.6
Lepidodendron aculeatum	UCM	N/A	PC	N/A	17.2A
Lepidodendron lycopodioides	UCM	P 221	PC	N/A	17.3
Lepidodendron obovatum	UCM	N/A	PC	N/A	17.2B
Lepidophloios laricinus	UCM	N/A	MSC	N/A	17.2D
Lepidophloios laricinus	UCM	N/A	MSC	N/A	17.2E
Lepidostrobus isp.	UCM	N/A	UF	UF 34365	17.4
Lycopoda[13]	UCM	N/A	ALMNH	PB2013.0014.0008	3.2
Neuralethopteris biformis	UCM	P 2	MSC	N/A	17.5
Sigillaria elegans	UCM	N/A	PC	N/A	17.2C
Raindrops	UCM	1024	MSC	MSC 26791	9.7

Appendix 2 Table Notes

Col. 1: The binomial name of the illustrated trace fossil, including the *Ichnogenus* (italics with capitalized first letter) and the *ichnospecies* (italics with lower case first letter); isp. means an ichnospecies has not yet been assigned.

Col. 2: UCM=Union Chapel Mine (Minkin Paleozoic Footprint Site). CVM=Crescent Valley Mine.

Col. 3: A number assigned to the specimen at a "track meet" or at the time it was first photographed; N/A means the specimen lacks this kind of number. P numbers refer to plant fossil specimens that were photographed at Track Meet 3 in 2001.

Col. 4: The institutional or other location of the specimen. ALMNH=Alabama Museum of Natural History; MSC=McWane Science Center. UF=University of Florida. PC=the specimen is in a private collection.

Col. 5: The institutional catalog number of the specimen. N/A means no number is available. Note that some specimens in institutions have not yet been assigned a formal catalog number.

Col. 6: The figure number of the specimen in this book.

1. Insect? or plant material?

2. Holotype

3. Abruptly ending trace

4. Interpretation uncertain; first specimen found at site

5. Reverse of UCM 1070

6. Counterimpression of 209

7. Largest track found

8. Holotype

9. Obstacle avoidance

10. Holotype

11 Holotype

12. Rare case with two tetrapod ichnotaxa on same horizon

13. "Devils Tower" trunk cast

Notes

Chapter 1

1. Letter from W. F. Cobb Sr. to Walter B. Jones, November 28, 1929, Geological Survey of Alabama archives, Tuscaloosa.

2. As recounted in Aldrich and Jones (1930).

3. Letter from W. F. Cobb Sr. to Walter B. Jones, December 27, 1929, Geological Survey of Alabama archives, Tuscaloosa.

4. Letter from W. B. Jones to W. F. Cobb Sr., January 3, 1930, Geological Survey of Alabama archives, Tuscaloosa.

5. Notes of George Gaylord Simpson, January 24, 1930, Geological Survey of Alabama archives, Tuscaloosa.

6. James Lamb, Gorden Bell, and Susan Henson originally formed the BPS to "recruit support for the Red Mountain Museum's program of salvage and research" (Gorden Bell, pers. comm. 2011).

7. Local resident John Southard explained the origin of the name Union Chapel: "When this area was settled there weren't enough of the different denominations of churches to justify building like a Baptist church, and a Methodist church, and different churches, so they said well let's just build a chapel, and we'll all use it. . . . They called it Union Chapel, and that's how the community as a whole got its name."

"Union Chapel" is a common church name in the United States. The church in Union Chapel, Alabama, is called the Union Chapel United Methodist Church, but an Internet search produces many churches around the country that either have the name "Union Chapel" or had it at one time. For example, the Flohrville United Methodist Church in Sykesville, Maryland, was once called "Union Chapel" because its founder, a German immigrant, wanted a nonsectarian church, one that would be a "union of all churches" (http://sites.google.com/site/centennialchurches/centennial-churches/flohrville-umc). This church was built in 1911, but the name "Union Chapel" goes back much further, to the United Kingdom. The Union Chapel Church in London was established in 1806 to "unite Christians of different denominations in religious worship and brotherly affection" (http://www.unionchapel.org.uk/history-of-union-chapel.html). Thus, Union Chapel is more than just a location in Alabama; it represents a concept that was popular at one time at many locations around the country and even the world.

8. The area is steeped in Civil War–era history that provides an interesting backdrop to our story. Considerable lore and legend exist about the time, near the end of the Civil War, when Union army troops passed through Alabama on their way to Georgia. The path these troops traveled was studied in detail by Rowland (1993), who provides evidence that Union troops passed very near the UCM area in late March 1865. On the way, they sometimes had to raid local farms and homes for food supplies, and there were skirmishes with local residents. But perhaps the most infamous incident was the detachment of the First Brigade of the First Division, under the command of Brig. Gen. John T. Croxton, to make a raid on Tuscaloosa and destroy the University of Alabama, the Alabama Military School, and other public buildings. This took place just a few days after the troops passed near the future mine site, and it left a scar on the city (including the burning of a library) that is still remembered (B. Windham, *Tuscaloosa News*, February 13, 2011).

9. "January 23, 2000 – Pennsylvanian Fossils, Walker Co, AL," Birmingham Paleontological Society, http://bps-al.org/trips/20000123.html.

10. Except for Rindsberg (1990).

Chapter 2

1. "Rocky Mountains," *Wikipedia*, http://en.wikipedia.org/wiki/Rocky_Mountains.

2. An example of this was found in eastern Illinois. An earthquake lowered the land level of a coal-age forest enough that a local river flooded it and buried it in sediment. The entire forest over 10 km² (4 mi²) was preserved standing. "Natural History Highlight: Four Square Miles of Carboniferous Forest Discovered," Smithsonian National Museum of Natural History, http://www.mnh.si.edu/highlight/riola/.

3. Limestone is sedimentary rock made up largely of the shells of organisms, such as seashells and the microscopic shells of foraminifera (shelled amoebas) and coccolithophores (floating algae).

4. The European series and stages do not exactly correspond to the North American periods. However, we use the European stages, such as Westphalian, to describe the Union Chapel Mine deposits because their fine subdivisions allow us to specify the age more precisely.

5. "*Appalachiosaurus montgomeriensis*," *Encyclopedia of Alabama*, http://www.encyclopediaofalabama.org/face/Article.jsp?id=h-2319.

6. For a summary of Alabama geology see "Geology of Alabama," *Encyclopedia of Alabama*, http://eoa.duc.auburn.edu/face/Article.jsp?id=h-1549.

7. "Bituminous Coal," *Wikipedia*, en.wikipedia.org/wiki/Bituminous_coal.

Chapter 3

1. Spencer Lucas, former curator of paleontology at the New Mexico Museum of Natural History and Science, noted how the Union Chapel Mine fossils changed the view of early reptiles. Bones of early reptiles are very rare, yet in Alabama tracks of early reptiles are common in some areas, proving that these animals were already abundant by the Late Carboniferous. In fact, sometime after Lucas visited the Minkin Site (in 2004), a group of British colleagues asked him to join them in writing a paper about the origin of reptiles. Among other things, they wanted to show that reptiles were absent from Carboniferous coal swamps. The Minkin Site fossils proved that this was not true.

2. "May 28, 2000 – Pennsylvanian Fossils, Walker Co, AL," Birmingham Paleontological Society, http://bps-al.org/trips/20000528.html.

3. The fossils from this rock, called the "Big Rock," are illustrated in Buta et al. (2005, plates 90–102). Several are shown in this book. All or most of the specimens that came from the rock are now held at the McWane Science Center.

4. An excellent schematic of lateral sequence walking is provided in plate 1 of Seilacher (2007). This walking technique is also diagrammed in steps in Clack (2012).

5. Seilacher (2007, plate 7) shows an example of this.

6. "Death trackways" have been found for invertebrates. Well-known cases of *Mesolimulus walchi* horseshoe crab body fossils lying at the ends of their trackways have been found in the Solnhofen Limestone in Germany. An example is shown at http://en.wikipedia.org/wiki/Mesolimulus.

7. "Photo in the News: Pre-Dino Amphibian Body Casts Found," *National Geographic News*, http://news.nationalgeographic.com/news/2007/10/071030-amphibian-fossil.html.

8. "*Apatosaurus*," *Wikipedia*, http://en.wikipedia.org/wiki/Apatosaurus.

Chapter 4

1. The earliest tetrapods had many toes (up to eight per foot). Until the beginning of the Carboniferous (about 359 million years ago), labyrinthodonts (which had anywhere from four to many toes on the manus) were the closest thing to what most specialists would call an amphibian. The oldest true amphibians (five or fewer toes per manus) were early Carboniferous. It is not known when the

amphibian manus lost its fifth toe for good, but it was probably long before the Westphalian.

2. For example, Falcon-Lang et al. (2010).

3. J. Lacefield found both reptile and amphibian tracks in the Fern Springs Mine northwest of Eldridge. A local landowner knowledgeable about mining in the area told him that the stratigraphic interval exposed in the Fern Springs Mine was near the Jagger seam. This is the same coal bed that was mined in Carbon Hill. Two other collectors (A. Allen and P. Atkinson) reported finding both reptile and amphibian tracks at another Black Warrior Basin mine, the Sugar Town Mine. The Minkin Site is thus not the only place where amphibian and reptile trackways co-occur in the Mary Lee coal zone. This makes the Black Warrior Basin even more significant to the study of early interactions between amphibians and reptiles than we knew. The Carbon Hill area appears to be an exception, however, as no amphibian tracks were found at the No. 11 Mine or the nearby Crescent Valley Mine (chapter 19).

Chapter 5

1. The Earth would also have some residual heat left over from its formation. In the case of the giant planets, this is the principal source of internal heat. Hamish Johnston, "Radioactive Decay Accounts for Half of Earth's Heat," Physicsworld.com, http://physicsworld.com/cws/article/news/2011/jul/19/radioactive-decay-accounts -for-half-of-earths-heat.

2. "Mantle Convection," *Wikipedia*, http://en.wikipedia.org/wiki/Mantle_convection.

3. A nice discussion of this is provided by Anne Egger, "Plates, Plate Boundaries, and Driving Forces," Visionlearning, http://www.visionlearning.com/library/module_viewer.php?mid=66.

4. Alfred Wegener, the man who first proposed "continental drift," the idea that later led to the full-fledged theory of plate tectonics, was ridiculed in his time (early twentieth century) because it was not clear how continents could really move. Now, however, plate tectonics is an accepted geological process that clearly shapes our world. Plate movements are measured in real time. Alabama once contained major active fault zones in the Earth's crust and was undoubtedly subject to earthquakes at a higher level than it is today. Mild quakes still occur in Alabama; they are relics from the tectonically driven upheaval of our distant past.

5. Pashin (2005) summarizes the history of Pennsylvanian deposition in Alabama as it relates to the Minkin Site. Because coal and natural gas are found in quantity in the Pottsville Formation (which contains the trackways at the Minkin Site), this unit has received a great deal of study.

6. C. R. Scotese, "The Paleomap Project," http://www.scotese.com.

7. J. Lacefield, pers. comm., 2011, used with permission.

8. Coauthor David has visited a hammock in South Florida. The floor of the hammock is only about half a meter higher than the surrounding marsh. But while the marsh is a wet plain covered with grasses and other relatively small plants, the hammock is a dense but tiny forest in which the trees are covered with bromeliads and Spanish moss. They could hardly look more different in life. Any preserved fossils from a hammock would be of different species than those found in marsh deposits. Yet the hammock is only hundreds of yards across and its boundary is sharp.

9. *"Lepidodendron," Wikipedia*, http://en.wikipedia.org/wiki/Lepidodendron.

10. Ibid.

11. William Sharpton, a geologist working for the National Coal Company, told us of an interesting but undocumented exception to this. He relates the following: "My first job out of college was with Bankhead Mining Co., based here in Jasper. The president was Frank Cobb, Jr., whose father Frank Sr. had been chief engineer of Galloway Coal Co. Frank Jr. told me stories his father told him about the tracks they frequently found in the roof of the Galloway mine. Of particular interest to me was one he told of a long series of smaller tracks that were met by a larger series of tracks [likely *Attenosaurus subulensis*] and where the two met just the larger series continued. I think it was Frank Sr. who contacted the geological survey folks about the tracks" (e-mail message, May 2013).

12. Coauthor David remembers seeing a herd of fiddler crabs on a Florida beach, and finding thousands of baby horseshoe crabs on the same beach on another visit.

Chapter 6

1. The concept of "today" is something we can experience only in our own solar system. All throughout the universe, events are happening "today," but we cannot know about these events until their light reaches us sometime in the distant future.

2. "Orion Nebula," *Wikipedia*, http://en.wikipedia.org/wiki/Orion_Nebula.

3. According to the NASA/IPAC Extragalactic Database (NED), http://ned-www.ipac.caltech.edu, based on an average line-of-sight speed of 6925 kilometers per second and a Hubble constant of 73 kilometers per second per megaparsec.

4. This graph is based on information from "Oort Constants," *Wikipedia*, http://en.wikipedia.org/wiki/Oort_constants; on new estimates of galactic constants from Ghez et al. (2008); and on equations for first-order departure from circular motion from Gilmore, King, and van der Kruit (1989).

5. The Big Dipper is not considered a full "constellation" but rather just a part of the constellation known as Ursa Major, or the "Big Bear." It is an asterism in the sense of being an easily recognizable pattern of stars.

6. Also known as Alpha Centauri.

7. "A Family Portrait of the Alpha Centauri System," ESO, http://www.eso.org/public/news/eso0307/.

8. "Main Sequence," *Wikipedia*, http://en.wikipedia.org/wiki/Main_sequence.

9. According to *Wikipedia*, the mass (in solar masses) for each bright Orion star is Rigel (21), Bellatrix (8–9), Saiph (15–16), Alnitak (33), Alnilam (40), Mintaka (20), and Betelgeuse (8–20); http://en.wikipedia.org/wiki/Rigel, http://en.wikipedia.org/wiki/Bellatrix, http://en.wikipedia.org/wiki/Saiph, http://en.wikipedia.org/wiki/Alnitak, http://en.wikipedia.org/wiki/Mintaka, http://en.wikipedia.org/wiki/Betelgeuse, and http://astropixels.com/stars/Alnilam-01/html

10. "Betelgeuse," *Wikipedia*, http://en.wikipedia.org/wiki/Betelgeuse.

11. "Polaris," *Wikipedia*, http://en.wikipedia.org/wiki/Polaris.

12. "Tycho (Crater)," *Wikipedia*, http://en.wikipedia.org/wiki/Tycho_%28crater%29.

13. "Astronomy Picture of the Day: Saturn's Ancient Rings," NASA, http://apod.nasa.gov/apod/ap071217.html.

14. "Tidal Acceleration," *Wikipedia*, http://en.wikipedia.org/wiki/Tidal_acceleration#Quantitative_description_of_the_Earth-Moon_case.

15. "Orbit of the Moon: Tidal Evolution," *Wikipedia*, http://en.wikipedia.org/wiki/Orbit_of_the_Moon#Tidal_evolution.

Chapter 7

1. "Carbon Hill, Alabama," *Wikipedia*, http://en.wikipedia.org/wiki/Carbon_Hill,_Alabama.

2. "Truman H. Aldrich," *Wikipedia*, http://en.wikipedia.org/wiki/Truman_H._Aldrich.

3. Walter B. Jones diaries, Hoole Special Collections, box 5118, folder 2, University of Alabama.

4. C. M. Whitson, pers. comm., 2012; Annual Report of Coal Mines, State of Alabama.

5. This map is dated March 12, 1935, which is likely the date the mine was permanently closed (C. M. Whitson, pers. comm. 2012).

6. *Coal Age* 35, no. 2 (1930).

7. *Coal Age* 33, no. 10 (1928).

8. Galloway Coal Company first appears at the location of the No. 11 Mine in the Annual Report of Coal Mines for 1913. Prior to this date, the mine was run by the Choctaw Coal and Mining Company, based in Memphis. This could not

have been for more than a year or two since it is absent from the 1910 issue (C. M. Whitson, pers. comm. 2012).

9. Folder on file at the Geological Survey of Alabama library with copyedited manuscript of Museum Paper 9.

10. "Left Tracks in 249,998,000 B.C. . . . Animals Roamed in Alabama," *Boston Herald*, January 29, 1930.

11. "Finds Florida Man of 20,000 Years Ago . . . Many Tracks Left 250,000,000 Years Ago Are Unearthed in Alabama," *New York Times*, January 29, 1930.

12. "Fossilized Record of the First Heavy-Weight Battle Found in Alabama," *Portsmouth (OH) Daily Times*, May 18, 1930.

13. "Fossilized Record of the First Heavy-Weight Battle Found in Alabama," *Ogden (UT) Standard-Examiner*, May 18, 1930.

14. Letter from F. N. Fisher, chairman of the Memphis Park Commission (and also president of the Galloway Coal Company) to George Gaylord Simpson, concerning an agreed-upon exchange of specimens, April 18, 1930, Geological Survey of Alabama Archives.

15. R. O'Leary, director of collections and archives, Division of Paleontology, American Museum of Natural History, pers. comm., October 23, 2012.

16. T. Braithwaite, registrar of collections, Memphis Pink Palace Museum, pers. comm., November 1, 2012.

17. M. Lockley points out that "the whole episode involving the American Museum's interest in the Alabama footprints is symptomatic of a mini-renaissance in fossil footprint research that took place in the 1930s. Simpson's visit is of historical interest because soon after his Alabama trip, the American Museum went in active search of fossil footprints in Texas, Colorado, Arizona and other places far from New York. Such is the potential of finds in far-flung locations to influence the thinking of leading paleontologists at major metropolitan museums, even in cases where the conscientious scientific follow up is rather sketchy" (pers. comm. 2013).

Chapter 8

1. A technique designed to enhance the sharper elements of an image. Specifically, the GNU Image Manipulation Program (GIMP) was used to convert the original digital JPEG image into a flexible image transport system (FITS) image. Then the Image Reduction and Analysis Facility (IRAF; iraf.noao.edu) was used to median-smooth this image, subtract it from the original, and then edit out the cracks using linear interpolation. The subtracted image was then contrast-enhanced to emphasize the tracks.

2. Martin and Pyenson (2005) present another example in which a small amphibian appears to have avoided a plant fragment.

3. During that first year after the initial discovery, coauthor David Kopaska-Merkel was working down the hall and around the corner from Dr. Charles C. Smith at the Geological Survey of Alabama. Dr. Smith is a paleontologist who knows a great deal about Alabama stratigraphy. He is now retired, but he used to drop everything to talk with a visitor. Steve Minkin stopped by the Geological Survey building twice a month after the Union Chapel discoveries to talk to Charlie about everything connected with the mine: what kinds of fossils BPS members were finding, how the trace fossils might have been made, new ideas about how to preserve the mine for the future, and what the mine site was like 300 million years ago when the trackways were formed. David can still see those two enthusiasts filling Charlie's small office with visions of Pennsylvanian landscapes.

Chapter 9

1. Jack Pashin is now at Oklahoma State University, Boone Pickens School of Geology.
2. New Acton Coal is currently owned by Robert Reed.
3. Don't climb a cliff if you are under eighteen, alone, not wearing a hard hat, or don't have experience and training in climbing rocks.
4. For example, Li et al. (2000).

Chapter 10

1. Permits are generally required to collect fossils from public land, but the state of Alabama places few restrictions on collecting fossils. The permission of the landowner is required to collect material from private land. Human remains and artifacts may not be dug out of the ground, although arrowheads may be collected from the surface. Nothing may be collected in caves except for scientific purposes. Finally, Alabama's state fossil, the Eocene whale *Basilosaurus cetoides*, may be collected, but specimens may not be removed from the state without the written permission of the governor. The federal government requires a permit to collect fossils on federal land, such as national parks and forests. However, most federal land in Alabama is relatively unfossiliferous. Andrew K. Rindsberg, "Laws about Fossils," Geological Survey of Alabama State Oil and Gas Board, http://www.gsa.state.al.us/gsa/fossil_laws.html.
2. Dendrites are mineral deposits, crystal growths that can resemble plant fossils. Further information may be found at "Pseudofossil," *Wikipedia*, http://en.wikipedia.org/wiki/Pseudofossil.
3. "Pelycosaur," *Wikipedia*, http://en.wikipedia.org/wiki/Pelycosaur.
4. At the Union Chapel Mine, we were able to drive our vehicles close enough to the rock piles to minimize the walking distance.

5. Regarding vertebrate tracks on public land, it would technically be illegal to collect specimens for a private collection in any case.

6. "Prehistoric Trackways National Monument," *Wikipedia*, http://en.wikipedia.org/wiki/Prehistoric_Trackways_National_Monument.

Chapter 11

1. Ron Buta, "The Photographic Trackway Database," http://www.alabamapaleo.org.

Chapter 12

1. For more information, see "Mine Reclamation," *Wikipedia*, http://en.wikipedia.org/wiki/Mine_reclamation.

2. This unfortunate split occurred because of procedural disagreements between longtime BPS members and newer members, and because of different views on the role the amateur group should play in Alabama paleontological studies.

3. Tony Martin of Emory University, Jerry MacDonald of the Paleozoic Trackways Project in New Mexico, and Hans-Dieter Sues, who at the time was president of the prestigious Society of Vertebrate Paleontology.

Chapter 15

1. Normal distributions can allow scientists to distinguish between fossils of different but similar species, between males and females, and even between different age groups of a single species. Any normally distributed groups, if different enough to separate on a graph, can be identified this way. Sometimes two overlapping normal distributions will yield a two-peaked graph in a set of size measurements of a sample containing members of both species (Müller and Walossek 1985). Even without complete separation of the two groups, a two-peaked distribution can prove the presence of two distinct groups.

2. This type of reptile has a temporal fenestra, or hole in the lower skull just behind the eye socket (Romer and Parsons 1985).

3. For example, Martin, Vazquez-Prokopec, and Page (2010).

Chapter 16

1. "Horseshoe Crab," *Wikipedia*, http://en.wikipedia.org/wiki/Horseshoe_crab.

2. "Myriapoda," *Wikipedia*, http://en.wikipedia.org/wiki/Myriapoda.

3. The "aps" in *T. apsorum* stands for Alabama Paleontological Society.

4. *A. repetitus* is mistakenly identified as a xiphosuran species by Buta et al. (2005).

5. Pashin (2005) argues that the depositional cycles in the upper Pottsville Formation each represent less than 500,000 years of time.

6. "Megasecoptera," *Wikipedia*, http://en.wikipedia.org/wiki/Megasecoptera.

7. "Odonatoptera," *Wikipedia*, http://en.wikipedia.org/wiki/Odonatoptera.

Chapter 18

1. See http://www.alabamapaleo.org.

Chapter 19

1. A rock layer may not be exactly the same age everywhere. Over time a particular environment may move, such as a beach-lagoon system gradually building out to sea. In the resulting deposit, the landward side would be a little older than the seaward side. However, the coastal mudflats and swamps that we call the *Cincosaurus* beds grew between two successive coal beds and would be *about* the same age everywhere. They would certainly be younger than rocks below the Jagger coal, and older than rocks above the Blue Creek.

2. The area covered by the CVM is mostly outside the boundary of the No. 11 Mine and is instead in an area where the 1935 No. 11 Mine map (which was used to make fig. 7.2) indicates that the Thomas Creek Coal Company operated. This may have been the company that left the observed cavities.

3. These pictures were brought to our attention two years after Buta's meeting with Donny Williams. William Sharpton, a geologist with the National Coal Company in Jasper, Alabama, told us that he had also seen the same vertebrate ceiling tracks that Williams saw. Most importantly, he recalled that these tracks were large, not like the typical *Cincosaurus cobbi* tracks found in the area. Sharpton brought us in contact with Anthony J. Edwards (a geologist and property manager for Regions Natural Resource Department in Birmingham), who had seen the same display and had taken the photographs in figure 19.1 at the time.

4. A few weeks after this first visit, another pit exposed a large, slightly deeper cavity than these others. This cavity was clearly related to the Jagger seam and although no tracks were seen in the ceiling, collapsed roof shale in the cavity yielded a *Cincosaurus cobbi* trackway nearly a meter long (Buta et al. 2013).

5. Old mining cavities such as those that were exposed at the CVM in 2008 are not, in general, safe to walk into. While it was possible to see tracks in the ceilings of some of these cavities, getting a closer look at them or retrieving specimens was dangerous because of the possibility of collapse. Donny Williams also mentioned the hazard of "black damp," where the air trapped for decades in a cavity is deficient in oxygen.

6. All CVM specimens are assigned running numbers in the manner of the Minkin Site "track meets."

7. The other two are "featherstitch" traces, where the animal left a trail resembling a type of embroidery stitch, and leveed traces, where the animal left a deeper body impression that appears "sandbagged," as in a levee.

8. "Trace Fossils of the Crescent Valley Surface Coal Mine, Carbon, Hill, Alabama," http://www.alabamapaleo.org.

9. The Massey Mine (MM) and the Lost Creek No. 7 Mine (LC7M) were brought to our attention by W. Sharpton, geologist with the National Coal Company.

Appendix 1

1. Some dates from notes by A. K. Rindsberg, "Union Chapel Mine: Events and Publications," unpublished.

Glossary

ALMNH Alabama Museum of Natural History, Tuscaloosa.

AMNH American Museum of Natural History, New York.

Amniote An organism producing eggs that can develop out of water because the embryos are enclosed in tough membranous sacs. At the time the *Cincosaurus* **beds** were deposited, the only amniotes were reptiles. Mammals and birds, descended from reptiles, are also amniotes.

Anthracosaur An extinct **tetrapod** physiologically and anatomically intermediate between amphibians and reptiles.

Appalachian Mountains An old mountain chain in eastern North America stretching from Alabama to the Atlantic provinces of Canada. The Appalachians are the product of three episodes of mountain building, but the last of these three episodes ended during the Pennsylvanian, before the first dinosaur evolved.

APS Alabama Paleontological Society. A group of amateur fossil collectors, and the driving force behind the documentation and preservation of the Union Chapel Mine. The group was established in 2002 as an offshoot of the Birmingham Paleontological Society (see **BPS**). Website: http://www.alabamapaleo.org.

Arborescent Having a treelike growth habit.

Arborichnus An invertebrate **trace fossil** common at the **UCM** but not found in the *Cincosaurus beds*. Made by an **arthropod** with at least eight legs.

Arenicolites longistriatus A *U*-shaped burrow. At the **UCM** these are attributed to dipteran (fly) larvae. See *Treptichnus apsorum*.

Arthropod Creatures with jointed legs. Arthropods evolved around the beginning of the Cambrian and have been the most successful group of animals on Earth. One group of arthropods, the insects, accounts for more species alive today than do all other groups of animals put together.

Attenosaurus subulensis A fossil **trackway** made by an animal that was likely a top predator on the **mudflats** during the coal age. These are the largest footprints found in the *Cincosaurus* **beds** and have been attributed to an **anthracosaur**.

Biramous Branching into two parts. Horseshoe crabs have biramous appendages, but insects do not.

BPS Birmingham Paleontological Society. A group of amateur fossil collectors. Founded in 1984, the BPS was the primary group involved in the Union Chapel Mine project from 2000 to 2002, after which the project was managed by the Alabama Paleontological Society (see **APS**). Website: http://bps-al.org.

Calamites Trunk fossil, often a cast, of a giant extinct horsetail. Modern horsetails are small and live in marshy areas. See also **sphenophyte**.

Carboniferous Period The period of Earth's history that in North America is called the Mississippian and the **Pennsylvanian**. The first reptiles evolved from amphibians during the Carboniferous, and a substantial fraction of the world's coal formed during this time.

Cincosaurus **beds** An informal rock unit, part of the **Pottsville Formation**, which contains **vertebrate trackways** at the Union Chapel Mine. The beds are named for the ichnospecies *Cincosaurus cobbi*, a vertebrate trackway. The beds also include a variety of other **trace fossils**. Part of the **Mary Lee coal zone**.

Cincosaurus cobbi The fossil **trackway** of a small animal that had five toes on each foot, with thumb-like digits pointing outward on the hind feet. Named after W. Frank Cobb Sr., the superintendent of the Galloway Coal Company No. 11 Mine, where the fossil was discovered.

Coal age See **Pennsylvanian Period**.

Coal seam A bed or layer of coal.

Coal zone A body of rock consisting of one or more **coal seams** and the intervening rocks (usually dominated by shale and sandstone).

Counterimpression Negative **impression**. Hardened sediment that, when soft, filled a footprint or other impression. The impression of a footprint is a low place on the top side of a rock; the equivalent counterimpression is a raised area on the bottom of a rock.

CVM Crescent Valley Mine, a surface coal mine located near Carbon Hill, Alabama. Active from 2008 to 2012 and owned by National Coal Company at that time, the mine was a rich source of coal-age **trace fossils**. The mine is of special interest because of its location on the northern edge of the Galloway Coal Company No. 11 Mine, where fossil tracks were first discovered in Alabama.

Ecosystem The organisms that live together on a certain part of the Earth and the climate, weather, and geology that combine to make living conditions more or less consistent throughout the area. For example, all the creatures living in the Mobile River **estuary**.

Epirelief Preserved on the top of a bed. Negative epirelief means indented into the top surface of a rock.

Estuary The broad mouth of a river where it meets the sea. Freshwater and seawater mix and tidal influence diminishes toward the land.

Fossiliferous Containing fossils.

Gait The way an animal walks (for example, fast, slow, bounding, trotting, loping).

Galloway No. 11 Mine An underground mine near Carbon Hill, Alabama, where fossil animal tracks were discovered in 1929. Active from 1912 to 1934.

Geologic column A graphical expression of major periods in Earth's history, arranged as a column with the most recent periods at the top and earlier periods below.

Hammock A drier and slightly elevated portion of a marsh. Vegetation is distinctly different from that of the surrounding marsh; hammocks are usually covered with trees.

Highwall Cliff created in a surface coal mine when overburden and coal are removed.

Holotype Particular specimen designated as representative of a species. Normally this is a mature, complete, and well-preserved specimen in which the species' diagnostic features are evident.

Hyporelief Preserved on the bottom of a bed. Positive hyporelief means protruding from the bottom surface of a rock.

Ichnofossil Trace fossil.

Ichnogenus A genus is a group of closely related species. An ichnogenus is a group of very similar **trace fossils**. The species that belong to a genus are united by common descent. The trace fossil species that belong to an ichnogenus are united only by similarity of form.

Ichnology The scientific study of **trace fossils**.

Ichnospecies A name assigned to a **trace fossil** within an ichnogenus.

Impression Footprint or other trace; a mark made directly on the substrate by an organism.

In situ In place. Found where it was made, in the case of a **trace fossil**.

Invertebrate An animal without a backbone. Not a fish, amphibian, reptile, mammal, or bird. An insect, a spider, a worm, or anything else that gets through life just fine without a spine.

Kouphichnium A **trace fossil** genus attributed to horseshoe crabs and characterized by marks made by **biramous** appendages.

Lagerstätte A mother lode. A body of rock, typically a rather small one, that is stuffed full of fossils. In the case of body fossils, a mother lode could imply mass death, such as might occur in a violent storm or earthquake, but in the case of **trace fossils**, it most likely implies fortuitous preservation in a small area.

Laminar Arranged in thin planes, as in the layers of some **sedimentary rock** units.

Lobe-finned fish Fish with stout bones in their fins that turned out to work well as primitive legs. They are the ancestors of all living **tetrapods**, including humans. A few species of lobe-finned fish survive today: coelacanths and lungfish. Most modern fish are ray-finned fish.

Lycopod Tall trees that grew during the **Westphalian**. They are called "scale trees" because their bark was covered with (leaf) scars that look like scales. Their tiny living relatives are called club mosses, although they have also been linked to small plants called quillworts. In life, the ancient lycopods would have had largely naked green trunks with small, narrow leaves clustered near the top.

Manus **Impression** of the forefoot of a **tetrapod**.

Mary Lee coal zone The coal zone in the **Pottsville Formation** that contains the *Cincosaurus* **beds**. The zone includes four major coal beds, named the Jagger, the Blue Creek, the Mary Lee, and the New Castle, and sits on a thick layer of sandstone.

Matthewichnus caudifer A **trackway** with four toes on the **manus** and five on the **pes**. Attributed to a small amphibian.

Megatracksite A horizon or restricted stratigraphic interval characterized by similar fossil track assemblages, exposed at multiple sites over an area of many square kilometers.

Merostome Either a horseshoe crab or a eurypterid (sea scorpion). A group of **arthropods** of which most members are extinct.

Monuran A member of a group of extinct wingless insects, makers of *Stiaria* and *Tonganoxichnus*.

Mudflat An area at the edge of a water body, perhaps in or near an **estuary,** where falling tides reduce water depth enough to expose a large muddy area to the air, allowing terrestrial animals to walk out into the mud to search for food.

Nanopus reidiae A common fossil **trackway** found at the **UCM** consisting of small footprints with five digits on the hind feet and four on the forefeet. Attributed to a **temnospondyl** amphibian.

Notalacerta A fossil **trackway** with five toes on both **manus** and **pes**. Attributed to an **amniote**.

Ouachita Mountains An ancient mountain chain, now mostly buried, that stretches westward from the southern end of the Appalachians in Alabama. The Ouachitas are still exposed in Oklahoma and Arkansas.

Pace The distance from one step to the next (for example, left front to right front or right rear to left rear).

Paleoenvironment An ancient environment: the living and nonliving components of any particular part of the world at some time in prehistory. In the **Westphalian A** paleoenvironment near Jasper one would have been well advised to put on a pair of waders.

Paleogeography The geography of the ancient world. The spatial arrangement of land and sea and of the various environments composing them at various times in the distant past. For example, 313 million years ago, central Walker County was an **estuary**.

Paleozoic Era The era of ancient life. That period in Earth's history stretching from the beginning of the Cambrian, 542 million years ago, to the end of the Permian, 251 million years ago. The **Westphalian A**, during which the *Cincosaurus* **beds** were deposited, was part of the Paleozoic.

Peat Partially decayed plant matter that gathers in a swamp environment, and that under pressure and heating can be converted into coal.

Pennsylvanian Period The period of time, ranging from 299 to 318 million years ago, during which the Pennsylvanian System, including rocks of **Westphalian** age, was deposited.

Pes Trace made by the hindfoot of a **tetrapod**.

Pottsville Formation A lower Pennsylvanian rock unit that contains the *Cincosaurus* **beds**. Named after Pottsville, Pennsylvania, the unit is made of **sedimentary rocks** deposited in a foreland basin created by a continental collision.

Pteridosperm **Seed fern**, an extinct kind of tree distantly related to ferns.

Reclamation The process of returning a surface coal mine to its premining state.

Reptile See **amniote**.

Sedimentary rock In most cases, a rock composed of particles that were transported to a common location and then deposited. Chemical rocks, like rock salt, are also considered to be sedimentary.

Seed fern An extinct group of trees with fernlike foliage. They were distantly related to ferns.

Sphenophyte A kind of plant with a straight, segmented stem and branches arranged in regularly spaced whorls. They were tree sized during the Paleozoic Era but are now very small. They are commonly known as horsetails. See also *Calamites*.

Stiaria An **invertebrate trackway** found at the **UCM**. Made by an animal that moved in short hops. Attributed to a wingless insect called a **monuran**.

Stratigraphy The science of the layering in sedimentary rock, the form of the layers and how they intergrade and correlate from place to place.

Stride The distance from one footprint to the next one made by the same foot.

Surface trackway A trail of footprints or marks exposed on the original surface where the marks were made. Often recognizable because a tail-drag mark is preserved. If well preserved, a surface trackway can provide the most accurate information on foot morphology and gait.

Temnospondyl A member of an extinct group of amphibians characterized by a special opening in the palate.

Tetrapod An animal that walks on four legs, such as most amphibians, reptiles, and mammals. Also, any animals descended from four-legged ancestors, such as snakes and birds.

Tonganoxichnus robledoensis An **invertebrate trace fossil** found at the **UCM** that appears as a series of small body **impressions** with tail marks impressed nearby. Attributed to the same **monuran** insect thought to have made *Stiaria*.

Trace fossil Evidence in the rock record of the previous existence of an animal that is not part of the animal itself, such as a footprint, crawl mark, jumping mark, tail drag, fin mark, burrow, or resting impression. Includes such nonbody remains as coprolites (fossil droppings).

Trackway A series of footprints (vertebrates) or walking marks (invertebrates) made by a single animal moving from one place to another.

Treptichnus apsorum A horizontal burrow consisting of segments oriented at angles to one another. Specimens at the **UCM** are attributed to dipteran (fly) larvae. See *Arenicolites longistriatus*.

UCM Union Chapel Mine. A small coal mine located approximately thirty-five miles west of Birmingham where fossil animal tracks were discovered by BPS members in late 1999 and early 2000. Owned and operated at that time by New Acton Coal Company, the mine was transferred to the state of Alabama in 2005 and renamed the Steven C. Minkin Paleozoic Footprint Site.

Undertrack A distortion formed *within* sediment when an animal steps on the sediment surface. Undertracks are footprints transmitted into layers of sediment that are not in actual contact with an animal's foot. Because they are already buried, undertracks typically have a higher preservation potential than do surface tracks. However, the deeper the undertrack, the less anatomical information about the foot is preserved.

Undichna A sinusoidal (wavy) fish-fin trace.

Vertebrate An animal with a backbone, including all **tetrapods** and fish.

Westphalian A Part of the Lower **Pennsylvanian Period** during which the *Cincosaurus* **beds** were deposited.

References

Aldrich, T. H., Sr., and W. B. Jones. 1930. *Footprints from the Coal Measures of Alabama*. Alabama Museum of Natural History, Museum Paper 9.

Anderson, A. 1976. "Fish Trails from the Early Permian of South Africa." *Palaeontology* 19:397–409.

Atkinson, T. P. 2005. "Arthropod Body Fossils from the Union Chapel Mine." In Buta, Rindsberg, and Kopaska-Merkel, *Pennsylvanian Footprints*, 169–76.

Atkinson, T. P., R. J. Buta, and D. C. Kopaska-Merkel. 2005. "Saving the Union Chapel Mine: How a Group of Determined Amateurs Teamed with Professionals to Save a World-Class Trackway Site in Alabama." In Buta, Rindsberg, and Kopaska-Merkel, *Pennsylvanian Footprints*, 191–200.

Beckemeyer, R. J., and M. S. Engel. 2011. "Upper Carboniferous Insects from the Pottsville Formation of Northern Alabama (Insecta: Ephemeropterida, Palaeodictyopterida, Odonatoptera)." *Scientific Papers, Natural History Museum, University of Kansas* 44:1–19.

Berman, D. S., A. C. Henrici, R. A. Kissel, S. S. Sumida, and T. Martens. 2004. "A New Diadectid (Diadectomorpha), *Orobates pabsti*, from the Early Permian of Central Germany." *Bulletin of the Carnegie Museum of Natural History* 35: 1–36.

Braddy, S. J. 2001. "Trackways: Arthropod Locomotion." In *Palaeobiology II*, edited by D. E. G. Briggs and P. R. Crowther, 389–93. Oxford: Blackwell.

Braddy, S. J., and D. E. G. Briggs. 2002. "New Lower Permian Nonmarine Arthropod Trace Fossils from New Mexico and South Africa." *Journal of Paleontology* 76:546–57.

Buatois, L. A., M. G. Mángano, C. G. Maples, and W. P. Lanier. 1998. "Ichnology of an Upper Carboniferous Fluvio-Estuarine Paleovalley: The Tonganoxie Sandstone, Buildex Quarry, Eastern Kansas, USA." *Journal of Paleontology* 72:152–80.

Buta, R. J., D. C. Kopaska-Merkel, A. K. Rindsberg, and A. J. Martin. 2005. "Atlas of Union Chapel Mine Invertebrate Trackways and Other Traces." In Buta, Rindsberg, and Kopaska-Merkel, *Pennsylvanian Footprints*, 207–76.

Buta, R. J., J. C. Pashin, N. J. Minter, and D. C. Kopaska-Merkel. 2013. "Ichnology and Stratigraphy of the Crescent Valley Mine: Evidence for a Carboniferous Megatracksite in Walker County, Alabama." In *The Carboniferous-Permian Transition*, edited by S. G. Lucas, W. J. Nelson, W. A. DiMichele, J. A. Spielmann, K. Krainer, J. E. Barrick, S. Elrick, and S. Voigt, 42–56. New Mexico Museum of Natural History and Science Bulletin 60.

Buta, R. J., A. K. Rindsberg, and D. C. Kopaska-Merkel. 2005. *Pennsylvanian Footprints in the Black Warrior Basin of Alabama*. Alabama Paleontological Society Monograph 1.

Butts, C. 1891. "Recently Discovered Foot-Prints of the Amphibian Age in the Upper Coal Measure Group of Kansas City, Missouri." *Kansas City Scientist* 5:17–19, 44.

———. 1926. "The Paleozoic Rocks." In *Geology of Alabama*, edited by G. I. Adams, C. Butts, L. W. Stephenson, and C. W. Cooke, 41–230. Alabama Geological Survey Special Report 14.

Caster, K. E. 1938. "A Restudy of the Tracks of *Paramphibius*." *Journal of Paleontology* 12:3–60.

Chesnut, D. R., D. Baird, J. H. Smith, and R. Q. Lewis. 1994. "Reptile Trackways from the Lee Formation (Lower Pennsylvanian) of South-Central Kentucky." *Journal of Paleontology* 68:154–58.

Clack, J. A. 2002. *Gaining Ground: The Origin and Evolution of Tetrapods.* 1st ed. Bloomington: Indiana University Press.

Clack, J. A. 2012. *Gaining Ground: The Origin and Evolution of Tetrapods.* 2nd ed. Bloomington: Indiana University Press.

Cleaves, A. W. 1983. "Carboniferous Terrigenous Clastic Facies, Hydrocarbon Producing Zones, and Sandstone Provenance, Northern Shelf of Black Warrior Basin." *Gulf Coast Association of Geological Societies Transactions* 33:41–53.

Colless, M. 2001. "Coma Cluster." In *Encyclopedia of Astronomy and Astrophysics*, edited by P. Murdin, article 2600. Bristol: Institute of Physics Publishing.

Collette, J. H., J. W. Hagadorn, and M. A. Lacelle. 2010. "Dead in Their Tracks: Cambrian Arthropods and Their Traces from Intertidal Sandstones of Quebec and Wisconsin." *Palaios* 25:475–86.

Davis, R. B., N. J. Minter, and S. J. Braddy. 2007. "The Neoichnology of Terrestrial Arthropods." *Palaeogeography, Palaeoclimatology, Palaeoecology* 255:284–307.

Dilcher, D. L., and T. A. Lott. 2005. "Atlas of Union Chapel Mine Fossil Plants." In Buta, Rindsberg, and Kopaska-Merkel, *Pennsylvanian Footprints*, 339–65.

Dilcher, D. L., T. A. Lott, and B. J. Axsmith. 2005. "Fossil Plants from the Union Chapel Mine, Alabama." In Buta, Rindsberg, and Kopaska-Merkel, *Pennsylvanian Footprints*, 153–68.

Dilcher, D. L., T. A. Lott, X. Wang, and Q. Wang. 2004. "A History of Tree

Canopies." In *Forest Canopies*, edited by M. D. Lowman and H. B. Rinker, 118–37. Burlington, MA: Elsevier Academic Press.

Donovan, S. K. 2010. "*Cruziana* and *Rusophycus*: Trace Fossils Produced by Trilobites . . . in Some Cases?" *Lethaia* 43:283–84.

Draganits, E., S. J. Braddy, and D. E. G. Briggs. 2001. "A Gondwanan Coastal Arthropod Ichnofauna from the Muth Formation (Lower Devonian, Northern India): Paleoenvironment and Tracemaker Behavior." *Palaios* 16:126–47.

Dudley, R. 1998. "Atmospheric Oxygen, Giant Paleozoic Insects and the Evolution of Aerial Locomotor Performance." *Journal of Experimental Biology* 201:1043–1050.

Eiseman, C., and N. Charney. 2010. *Tracks and Sign of Insects and Other Invertebrates*. Mechanicsburg, PA: Stackpole Books.

Evenhuis, N. L. 2004. *Catalogue of the Fossil Flies of the World (Insecta: Diptera)*. Leiden, Netherlands: Backhuys Publishers.

Falcon-Lang, H. J., M. J. Benton, and M. Stimson. 2007. "Ecology of Earliest Reptiles Inferred from Basal Pennsylvanian Trackways." *Journal of the Geological Society, London* 164:1113–18.

Falcon-Lang, H. J., M. R. Gibling, M. J. Benton, R. F. Miller, and A. R. Bashforth. 2010. "Diverse Tetrapod Trackways in the Lower Pennsylvanian Tynemouth Creek Formation, near St. Martins, Southern New Brunswick, Canada." *Palaeogeography, Palaeoclimatology, Palaeoecology* 296:1–13.

Ferm, J. C., R. Ehrlich, and T. L. Neathery. 1967. *A Field Guide to Carboniferous Detrital Rocks in Northern Alabama*. Geological Society of America Coal Division Guidebook.

Gardner, J. A. 1933. "Memorial of Truman Heminway Aldrich [1848–1932]." *Geological Society of America Bulletin* 44:301–7.

Gastaldo, R. A., T. M. Demko, and Y. Liu. 1990. "Carboniferous Coastal Environments and Paleocommunities of the Mary Lee Coal Zone, Marion and Walker Counties, Alabama. A Guidebook for Field Trip VI, 39th Annual Meeting, Southeastern Section of the Geological Society of America." *Tuscaloosa: Geological Survey of Alabama* 41–54.

Gastaldo, R. A., T. M. Demko, and Y. Liu. 1993. "Application of Sequence and Genetic Stratigraphic Concepts to Carboniferous Coal-Bearing Strata: An Example from the Black Warrior Basin, USA." *Geologische Rundschau* 82:212–26.

Gevers, T. W., L. A. Frakes, L. N. Edwards, and J. E. Marzolf. 1971. "Trace Fossils from the Lower Beacon Sediments (Devonian), Darwin Mountains, Southern Victoria Land, Antarctica." *Journal of Paleontology* 45:81–94.

Ghez, A. M., S. Salim, N. N. Weinberg, J. R. Lu, T. Do, J. K. Dunn, K. Matthews, et al. 2008. "Measuring Distance and Properties of the Milky Way's

Central Supermassive Black Hole with Stellar Orbits." *Astrophysical Journal* 689:1044–62.

Gilmore, G., I. King, and P. van der Kruit. 1989. *The Milky Way as a Galaxy: 19th Advanced Course of the Swiss Society of Astronomy and Astrophysics.* Geneva: Geneva Observatory.

Hall, J. 1852. *Palaeontology of New York, Volume II, Containing Descriptions of the Organic Remains of the Lower Middle Division of the New-York System (Equivalent in Part to the Middle Silurian Rocks of Europe).* Albany: Charles van Benthuysen.

Harrington, H. J., et al. 1959. *Treatise on Invertebrate Paleontology.* Part O, Arthropoda 1. Geological Society of America.

Haubold, H. 1971. "Ichnia Amphibiorum et Reptiliorum fossilium." *Encyclopedia of Paleoherpetology* 8:1–124.

———. 1998. "The Early Permian Tetrapod Ichnofauna of Tambach, the Changing Concepts in Ichnotaxonomy." *Hallesches Jahrbuch für Geowissenschaften* B20:1–16.

Haubold, H., A. Allen, T. P. Atkinson, R. J. Buta, J. A. Lacefield, S. C. Minkin, and B. A. Relihan. 2005a. "Interpretation of the Tetrapod Footprints from the Early Pennsylvanian of Alabama." In Buta, Rindsberg, and Kopaska-Merkel, *Pennsylvanian Footprints,* 75–111.

Haubold, H., R. J. Buta, A. K. Rindsberg, and D. C. Kopaska-Merkel. 2005b. "Atlas of Union Chapel Mine Vertebrate Trackways and Swimming Traces." In Buta, Rindsberg, and Kopaska-Merkel, *Pennsylvanian Footprints,* 207–76.

Haubold, H., and S. G. Lucas. 2001. "Early Permian Tetrapod Tracks: Preservation, Taxonomy, and Euramerican Distribution." *Monografie di Natura Bresciana* 25:347–54.

Hunt, A. P., S. G. Lucas, and M. G. Lockley. 2004. "Large Pelycosaur Footprints from the Lower Pennsylvanian of Alabama, USA." *Ichnos* 11:39–44.

Hunt, A. P., S. G. Lucas, and N. D. Pyenson. 2005. "The Significance of the Union Chapel Mine Site: A Lower Pennsylvanian (Westphalian A) Ichnological Konzentrat-Lagerstätte, Alabama, USA." In Buta, Rindsberg, and Kopaska-Merkel, *Pennsylvanian Footprints,* 3–14.

Keighley, D. G., J. H. Calder, A. F. Park, R. K. Pickerill, J. W. F. Waldron, H. J. Falcon-Lang, and M. J. Benton. 2008. "Discussion on Ecology of Earliest Reptiles Inferred from Basal Pennsylvanian Trackways," *Journal,* Vol. 164, 2007, 1113–1118.

Knecht, R. J., M. S. Engel, and J. S. Benner. 2011. "Late Carboniferous Paleoichnology Reveals the Oldest Full-Body Impression of a Flying Insect." *Proceedings of the National Academy of Sciences* 108(16): 6515.

Kohl, M. S., and J. R. Bryan. 1994. "A New Middle Pennsylvanian (Westphalian)

Amphibian Trackway from the Cross Mountain Formation, East Tennessee Cumberlands." *Journal of Paleontology* 68:655–63.

Kvale, E. P. 2006. "The Origin of Neap-Spring Tidal Cycles." *Marine Geology* 235:5–18.

Lacefield, J. A. 2000. *Lost Worlds in Alabama Rocks: A Guide to the State's Ancient Life and Landscape.* Birmingham: Alabama Geological Society.

Lacefield, J. A., and B. A. Relihan. 2005. "The Significance of the Union Chapel Mine Project to Alabama Paleontology." In Buta, Rindsberg, and Kopaska-Merkel, *Pennsylvanian Footprints,* 201–4.

Li, C., P. Wang, D. Fan, B. Dang, and T. Li. 2000. "Open-Coast Intertidal Deposits and the Preservation Potential of Individual Laminae: A Case Study from East-Central China." *Sedimentology* 47:1039–51.

Liu, Y., and R. A. Gastaldo. 1992. "Characteristics and Provenance of Log-Transported Gravels in a Carboniferous Channel Deposit." *Journal of Sedimentary Petrology* 62:1071–83.

Lockley, M. G., and A. Hunt. 1995. *Dinosaur Tracks and Other Fossil Footprints of the Western United States.* New York: Columbia University Press.

Lockley, M. G., and C. Meyer. 2000. *Dinosaur Tracks and Other Fossil Footprints of Europe.* New York: Columbia University Press.

Lockley, M. G., and J. Peterson. 2002. *A Guide to the Fossil Footprints of the World.* Denver: Lockley-Peterson.

Lucas, S. G., and A. B. Heckert, eds. 1995. *Early Permian Footprints and Facies.* New Mexico Museum of Natural History and Science Bulletin 6, Albuquerque.

Lucas, S. G., and A. J. Lerner. 2005. "Lower Pennsylvanian Invertebrate Ichnofossils from the Union Chapel Mine, Alabama: A Preliminary Assessment." In Buta, Rindsberg, and Kopaska-Merkel, *Pennsylvanian Footprints,* 147–52.

MacDonald, J. 1992. "Footprints from the Dawn of Time." *Science Probe,* July, 33–43.

———. 1994. *Earth's First Steps: Tracking Life before the Dinosaurs.* Boulder: Johnson Printing.

———. 1995. "History of the Discovery of Fossil Footprints in Southern New Mexico." In *Early Permian Footprints and Facies,* New Mexico Museum of Natural History and Science Bulletin 6, Albuquerque.

Mángano, M. G., L. A. Buatois, C. G. Maples, and W. P. Lanier. 1997. "*Tonganoxichnus,* a New Insect Trace from the Upper Carboniferous of Eastern Kansas." *Lethaia* 30:113–25.

Martin, A. J., and N. D. Pyenson. 2005. "Behavioral Significance of Trace Fossils from the Union Chapel Site." In Buta, Rindsberg, and Kopaska-Merkel, *Pennsylvanian Footprints,* 59–73.

Martin, A. J., G. M. Vazquez-Prokopec, and M. Page. 2010. "First Known Feeding Trace of the Eocene Bottom-Dwelling Fish *Notogoneus osculus* and Its Paleontological Significance." *PLOS One* 5, e10420, doi:10.1371/journal.pone.0010420.

McCalley, H. 1900. *Report on the Warrior Coal Basin*. Alabama Geological Survey Special Report 10.

Minter, N. J., and S. J. Braddy. 2006. "Walking and Jumping with Paleozoic Apterygote Insects." *Palaeontology* 49:827–35.

———. 2009. *Ichnology of an Early Permian Intertidal Flat: The Robledo Mountains Formation of New Mexico, USA*. Special Papers in Palaeontology 82.

Müller, K. J., and D. Walossek. 1985. "Skaracarida, a New Order of Crustacea from the Upper Cambrian of Västergötland, Sweden." *Fossils and Strata* 17:1–65.

Murie, O. J., and M. Elbroch. 2005. *A Field Guide to Animal Tracks*. Peterson Field Guides. Boston: Houghton Mifflin.

Nopsca, F. B. 1923. "Die Familien der Reptilien." *Fortschritte der Geologie und Paläontologie* 2:1–210.

Pashin, J. C. 1993. "Tectonics, Paleogeography, and Paleoclimatology of the Kaskaskia Sequence in the Black Warrior Basin of Alabama." In *New Perspectives on the Mississippian System of Alabama*, edited by J. Pashin, 1–28. Alabama Geological Society 30th Annual Field Trip Guidebook.

———. 1998. "Stratigraphy and Structure of Coalbed Methane Reservoirs in the United States: An Overview." *International Journal of Coal Geology* 35:207–38.

———. 2005. "Pottsville Stratigraphy and the Union Chapel Lagerstätte." In Buta, Rindsberg, and Kopaska-Merkel, *Pennsylvanian Footprints*, 39–58.

Pashin, J. C., R. E. Carroll, R. L. Barnett, and M. A. Beg. 1995. *Geology and Coal Resources of the Cahaba Coal Field*. Alabama Geological Survey Bulletin 163.

Pashin, J. C., W. E. Ward II, R. B. Winston, R. V. Chandler, D. E. Bolin, K. E. Richter, W. E. Osborne, and J. C. Sarnecki. 1991. *Regional Analysis of the Black Creek-Cobb Coalbed-Methane Target Interval, Black Warrior Basin, Alabama*. Alabama Geological Survey Bulletin 145.

Rindsberg, A. K. 1990. "Fresh Water to Marine Trace Fossils of the Mary Lee Coal Zone and Overlying Strata (Westphalian A) Pottsville Formation of Northern Alabama." In *Carboniferous Coastal Environments and Paleocommunities of the Mary Lee Coal Zone, Marion and Walker Counties, Alabama*, edited by R. A. Gastaldo, T. M. Demko, and Y. Liu, 82–92. Guidebook for Field Trip VI, 39th Annual Meeting, Southeastern Section, Geological Society of America.

———. 2005. "Gas-Escape Structures and Their Paleoenvironmental Significance at the Steven C. Minkin Paleozoic Footprint Site (Early Pennsylvanian, Alabama)." In Buta, Rindsberg, and Kopaska-Merkel, *Pennsylvanian Footprints*, 177–83.

Rindsberg, A. K., and D. C. Kopaska-Merkel. 2005. "*Treptichnus* and *Arenicolites* from the Steven C. Minkin Paleozoic Footprint Site (Langsettian, Alabama, USA)." In Buta, Rindsberg, and Kopaska-Merkel, *Pennsylvanian Footprints*, 121–42.

Romano, M., and B. Meléndez. 1985. "An Arthropod (Merostome) Ichnocoenosis from the Carboniferous of Northwest Spain." *Ninth International Geological Congress, Urbana, Illinois* 5:317–25.

Romer, A. S., and T. S. Parsons. 1985. *The Vertebrate Body*. 6th ed. Philadelphia: Saunders.

Rowland, A. 1993. *Wilson's Raid: Union Troops in Walker County, March and April 1865*. Jasper, AL: Walker College.

Scheckler, S. E. 2001. "Afforestation: The First Forests." In *Palaeobiology II*, edited by D. E. G. Briggs and P. R. Crowther, 67–71. Oxford: Blackwell.

Seilacher, A. 2007. *Trace Fossil Analysis*. Berlin: Springer.

Shubin, N. 2009. *Your Inner Fish: A Journey into the 3.5-Billion-Year History of the Human Body*. New York: Vintage Press.

Simpson, G. G. 1987. *Simple Curiosity: Letters from George Gaylord Simpson to Family, 1921–1970*. Edited by L. F. Laporte. Berkeley: University of California Press.

Smith, J. 1909. *Upland Fauna of the Old Red Sandstone Formation of Carrick, Ayrshire*. Kilwinning, UK: A. W. Cross.

Snodgrass, R. E. 1930. *Insects: Their Ways and Means of Living*. Paderborn, Germany: Salzwasser Verlag.

Soler-Gijón, R., and J. J. Moratalla. 2001. "Fish and Tetrapod Trace Fossils from the Upper Carboniferous of Puertollano, Spain." *Palaeogeography, Palaeoclimatology, Palaeoecology* 171:1–28.

Stanley, S. M. 2009. *Earth System History*. 3rd ed. New York: W. H. Freeman.

Thomas, W. A. 1988. "The Black Warrior Basin." In *Sedimentary Cover—North American Craton: U.S.*, edited by L. L. Sloss, D-2, 471–92. Geology of North America series, Geological Society of America.

Turek, V. 1989. "Fish and Amphibian Trace Fossils from Westphalian Sediments of Bohemia." *Palaeontology* 32:623–43.

Twenhofel, W. H., and R. R. Shrock. 1935. *Invertebrate Paleontology*. New York: McGraw-Hill.

Uchman, A. 2005. "*Treptichnus*-like Traces by Insect Larvae (Diptera: Chironomidae, Tipulidae)." In Buta, Rindsberg, and Kopaska-Merkel, *Pennsylvanian Footprints*, 143–46.

Voigt, S., D. S. Berman, and A. C. Henrici. 2007. "First Well-Established

Track-Trackmaker Association of Paleozoic Tetrapods Based on *Ichniotherium* Trackways and Diadectid Skeletons from the Lower Permian of Germany." *Journal of Vertebrate Paleontology* 27:553–570.

Wielen, R. 1985. "Dynamics of Open Star Clusters." In *Dynamics of Star Clusters*, edited by J. Goodman and P. Hut, 449–60, IAU Symposium 113. Dordrecht, Netherlands: Reidel.

Winston, R. B. 1991. *A Structurally Preserved, Lower Pennsylvanian Flora from the New Castle Coal Bed of Alabama.* Alabama Geological Survey, Circular 157.

About the Authors

Dr. Ronald J. Buta graduated from the Baltimore Polytechnic Institute in 1970 and received a PhD in astronomy from the University of Texas at Austin in 1984. His primary research interests are the morphology and dynamics of galaxies. Buta became interested in astronomy at the age of thirteen after a chance encounter with a beautiful clear night over Baltimore in the spring of 1965. He did not connect with paleontology until the early 1970s, when he was an undergraduate student at Case Western Reserve University in Cleveland, Ohio. There, he often visited the Cleveland Museum of Natural History, which was near campus. His interest in paleontology nevertheless remained dormant until he became a faculty member in the Department of Physics and Astronomy at the University of Alabama in 1989. In the mid-1990s, seeking a way to make introductory astronomy more relevant to students (particularly the part of the course concerned with processes that mold terrestrial planets), Buta took an interest in the natural history of Alabama. In 1997, coauthor Kopaska-Merkel introduced Buta to fossil collecting in Alabama, which eventually led to Buta's association with the BPS and the APS.

Dr. David C. Kopaska-Merkel was fascinated by fossils as a young child. His father was an amateur but very serious naturalist. He and his father visited fossil localities, mineral localities, and the homes of rare mountain plants pretty often. David collected diabase boulders (diabase is an igneous rock) and quartz crystals on his way home from school, and also fossil shark teeth and many other kinds of fossils whenever he could. He eventually became a professional paleontologist. He received a PhD in geology from the University of Kansas in 1983. He was first exposed to trace fossils when he discovered starfish resting traces at a well-known Kansas locality. He donated those specimens to the University of Kansas paleontological museum. Years later, David visited a Mississippian locality in northern Alabama where trilobite trackways and resting traces were well preserved and abundant. He had never collected or studied fossil vertebrate trackways before visiting the Union Chapel Mine.

Illustration Credits

Figure 1.1: Photo by Ron Buta

Figure 1.2: *Top left*, photo courtesy of Ashley Allen, Alabama Paleontological Society (APS); *top*, *middle*, and *bottom right*, photos by Ron Buta, courtesy of the APS

Figure 2.1*A*: Chart prepared by David Kopaska-Merkel

Figure 2.1*B*: Chart prepared by David Kopaska-Merkel

Figure 2.2*A*: Photo by Ron Buta

Figure 2.2*B*: Photo by Ron Buta

Figure 3.1: *Top*, photo courtesy of David S. Berman, Carnegie Museum of Natural History; *bottom*, photo by Ron Buta

Figure 3.2: Photo by Ron Buta

Figure 3.3: Photos by Ron Buta

Figure 3.4: Photo by Ron Buta

Figure 3.5: Illustration by Sue Blackshear; courtesy of the APS

Figure 3.6: Photo by Ron Buta

Figure 3.7: Specimen source: Crescent Valley Mine; photos by Ron Buta

Figure 3.8: Photo by Ron Buta

Figure 3.9: Photo and drawing by Ron Buta

Figure 4.1: *Left*, Aldrich and Jones (1930), public domain; *right*, courtesy of Haubold et al. (2005a)

Figure 4.2: Drawing by Zina Deretsky, courtesy of the National Science Foundation

Figure 4.3: Drawings courtesy of Jennifer A. Clack

Figure 4.4: Compiled by Sandy Ebersole, Alabama Geological Survey, public domain

Figure 5.1: Drawing by Jack C. Pashin, APS, reprinted with permission

Figure 5.2: Graphic courtesy of Jim Lacefield

Figure 5.3: Graphic courtesy of Jim Lacefield

Figure 5.4: Courtesy of the Field Museum, Chicago, Illinois, GEO85637c

Figure 5.5: Graphic courtesy of Jim Lacefield

Figure 6.1: Courtesy of Dean C. Rowe

Figure 6.2: Sloan Digital Sky Survey Collaboration, http://www.sdss.org

Figure 6.3: Sloan Digital Sky Survey Collaboration, http://www.sdss.org

Figure 6.4: Drawing by Ron Buta

Figure 6.5: Drawing by Ron Buta

Figure 6.6: Courtesy of ESA/Hubble Heritage

Figure 7.1: *Top left*, courtesy of Birmingham, Alabama, Public Library Archives, Portrait Collection; *top right*, courtesy of the Geological Survey of Alabama Archives; *bottom left*, portrait courtesy of W. Frank Cobb III; *bottom right*, courtesy of Birmingham, Alabama Public Library Archives, Portrait Collection

Figure 7.2: Drawing by Ron Buta

Figure 7.3: Taken from Aldrich and Jones (1930); photo by Dr. R. S. Hodges, Alabama Museum of Natural History

Figure 8.1: Photo by Ron Buta

Figure 8.2: Courtesy of Larry A. Herr

Figure 8.3: Photo by Ron Buta

Figure 8.4: Photo by Ron Buta

Figure 8.5: (*A*) photo by Ron Buta; (*B*) photo by T. Prescott Atkinson, APS

Figure 8.6: Photo by Ron Buta

Figure 8.7: Photos by Ron Buta

Figure 8.8: Photo by Ron Buta

Figure 8.9: Artwork by Sue Blackshear superimposed on photo by Ron Buta

Figure 8.10: Photos by Ron Buta

Figure 8.11: Photo by Ron Buta

Figure 8.12: Photos by Ron Buta

Figure 9.1: Photo by Ron Buta

Figure 9.2*A*: Drawing by Jack C. Pashin, 2005, APS; reprinted with permission

Figure 9.2*B*: Drawing by Jack C. Pashin, 2005, APS; reprinted with permission

Figure 9.3: Photos by Jack C. Pashin

Figure 9.4: Courtesy of S. G. Lucas and M. Stimson

Figure 9.5: *Top*, photo by Steven C. Minkin; *bottom*, photo by Andrew K. Rindsberg

Figure 9.6: Photo courtesy of J. K. Bartley and B. C. Bartley, unpublished results

Figure 9.7: Photo by T. Prescott Atkinson, APS

Figure 9.8: Drawing by Ron Buta

Figure 10.1: *Top*, photo by Jerry MacDonald; *bottom*, photo by Pearl MacDonald. Both photographs are courtesy of the Paleozoic Trackways Project Archives

Figure 11.1: Photo by Ron Buta

Figure 11.2: Photo by Ron Buta

Figure 12.1: Photo by Ron Buta

Figure 12.2: Photo by Ron Buta

Figure 12.3: Photo by T. Prescott Atkinson

Figure 13.1: *Top left*, photo by Ron Buta; *top right*, photo by Ron Buta; *bottom*, photo by Andrew K. Rindsberg

Figure 13.2: Photo by Ron Buta

Figure 14.1: Photo by Ron Buta

Figure 14.2: Drawing courtesy of H. Haubold, hartmut.haubold@geo.uni-halle.de

Figure 15.1: Photos by Ron Buta

Figure 15.2: Illustration by Sue Blackshear, courtesy of APS

Figure 15.3: Courtesy of H. Haubold, hartmut.haubold@geo.uni-halle.de

Figure 15.4: Illustration by Sue Blackshear, courtesy of APS

Figure 15.5: Courtesy of H. Haubold, hartmut.haubold@geo.uni-halle.de

Figure 15.6: Illustration by Sue Blackshear, courtesy of APS

Figure 15.7: (*A*) Courtesy of H. Haubold, reprinted with permission; (*B*) photo by Ron Buta; (*C*) photo by Ron Buta

Figure 15.8: Illustration by Sue Blackshear, courtesy of APS

Figure 15.9: Photos by Ron Buta

Figure 15.10: Illustration by Sue Blackshear, courtesy of APS

Figure 15.11: Photos by Ron Buta

Figure 15.12: Photos by T. Prescott Atkinson, with adaptations by Ron Buta

Figure 15.13: Illustration by Sue Blackshear, courtesy of APS

Figure 15.14: (*A*) Courtesy of H. Haubold, reprinted with permission; photos (*B*) and (*C*) by Ron Buta

Figure 16.1: Photo (*A*) courtesy of S. Lucas and M. Stimson; photos (*B*) and (*C*) by Ron Buta

Figure 16.2: Illustration by Sue Blackshear, courtesy of APS

Figure 16.3: Photos by Ron Buta

Figure 16.4: Photos (*A*) and (*B*) by Ron Buta; (*C*) courtesy of S. Lucas and M. Stimson

Figure 16.5: Illustration by Sue Blackshear, courtesy of APS

Figure 16.6: Photos (*A*) through (*C*) courtesy of S. Lucas and M. Stimson; photo (*D*) courtesy of Larry A. Herr

Figure 16.7: Illustration by Sue Blackshear, courtesy of APS

Figure 16.8: Photo (*A*) courtesy of S. Lucas and M. Stimson; photos (*B*) and (*C*) by Ron Buta

Figure 16.9: Illustration by Sue Blackshear, courtesy of APS

Figure 16.10*A*: Photos by Ron Buta

Figure 16.10*B*: *Top*, photo by Ron Buta; *bottom*, courtesy of S. Lucas and M. Stimson

Figure 16.10*C*: *Top*, photo courtesy of S. Lucas and M. Stimson; *bottom*, photo by Ron Buta

Figure 16.11*A*: Photos by Ron Buta

Figure 16.11*B*: Photos by Ron Buta

Figure 16.11*C*: Photos by Ron Buta

Figure 16.12*A*: Photos by Ron Buta

Figure 16.12*B*: Photos by T. Prescott Atkinson, APS

Figure 16.13: Photos courtesy of S. Lucas and M. Stimson

Figure 16.14: Photos (*A*) and (*B*) courtesy of S. Lucas and M. Stimson; photos (*C*) and (*D*) by Ron Buta

Figure 16.15*A*: *Top*, photo by Ron Buta; *bottom*, photo by T. Prescott Atkinson

Figure 16.15*B*: Photo by Ron Buta

Figure 16.15*C*: Photos by Ron Buta

Figure 17.1: Taken from Pashin (2005), APS

Figure 17.2: All photos by Ron Buta

Figure 17.3: *Top*, adapted from Dilcher, Lott, and Axsmith (2005), APS; *bottom*, photo by Ron Buta

Figure 17.4: Adapted from Dilcher, Lott, and Axsmith (2005), APS

Figure 17.5: Photo by Ron Buta

Figure 17.6: All photos courtesy of Dilcher and Lott (2005) and Dilcher, Lott, and Axsmith (2005); reprinted with permission

Figure 17.7: Adapted from Dilcher, Lott, and Axsmith (2005)

Figure 18.1: Photo by Ron Buta

Figure 19.1: All photos courtesy of Tony Edwards

Figure 19.2: Photo by Ron Buta

Figure 19.3: Photo by Ron Buta

Figure 19.4: Photo by Ron Buta

Figure 19.5: Photo by Ron Buta

Figure 19.6: Photo by Ron Buta

Figure 19.7: Photo by Ron Buta

Figure 19.8: Photo by Ron Buta

Figure 19.9: Photo by Ron Buta

Figure 19.10: Compiled and drawn by Ron Buta

Index

Aderholt, Robert, Alabama Congressman, 4th district, 160, 162–5, 167–8, 171, 173, 253, 275–6

Agaeoleptoptera uniotempla, 234–5, 279

Alabama Department of Conservation and Natural Resources, vii, 167, 171, 173, 276

Alabama Geological Society, iv, 116, 182, 274, 309–10

Alabama Geological Survey. *See* Geological Survey of Alabama

Alabama Museum of Natural History, iv, 1–2, 4, 49, 90–1, 101, 141, 146–7, 151, 158–9, 162, 166, 266, 274–5, 277, 283, 297

Alabama Paleontological Society, iv, vii, viii, xv, xviii, 118, 132, 165, 170, 174–5, 275, 293, 297–8

Alabama Surface Mining Commission, 155, 160, 166–71,173, 175, 275–6

Aldrich, Truman H., vii, xiii, 2, 4, 11, 14, 22, 50, 90–7, 121, 136, 157–8, 161–2, 182, 186, 188, 198, 256–7, 264, 273, 285, 290

Allen, Ashley, 5–12, 14, 22, 24, 38, 191, 103, 105, 116, 118, 142, 145, 155, 173, 178, 180, 182, 253, 274, 288

ALMNH. *See* Alabama Museum of Natural History

amniote, 50, 53–6, 181, 184, 186–8, 190, 192, 196, 198, 297, 301, 302

amphibian, iv, xii, 2–4, 6, 8–9, 22, 25–8, 30–1, 42, 45, 47–51, 53–6, 73, 75, 94, 98, 103, 107, 109–10, 129, 142, 176, 181, 184–6, 188, 190, 192, 194, 196–9, 205–6, 216, 237, 264–6, 287–8, 291, 297–8, 300–2, 306, 309, 311

amphibian-like reptile. *See* reptile-like amphibian

anal fin, 52, 201–2

Anniedarwinia alabamaensis, 106, 234–5, 279

Anniston Museum of Natural History, 101, 148–9, 151, 274

anthracosaur: 53–6, 75, 181, 196–7, 297–8. *See also Attenosaurus subulensis*

Apatosaurus, 47

Appalachian Mountains, 58–60, 68, 88, 297

APS. *See* Alabama Paleontological Society

Arborichnus repetitus, 225–8, 231, 279, 293

Arenicolites longistriatus, 223–4, 279, 297, 303

positive hyporelief, 26, 34, 36, 41, 51, 94, 110, 203, 224–7, 229, 251, 258, 299

Pottsville Formation, vii, 21, 49, 65–7, 74–5, 118–9, 129, 131, 186, 192, 288, 294, 298, 300, 302

preservation bias, 44

primary track, 35, 125, 188

pteridosperms, 49, 71, 239, 246

Pyenson, Nicholas R., 143, 163, 180

quillwort, 72, 300

raindrop impressions, 126–8

Red Mountain Museum, 4, 273, 285

Reid, Dolores, vii, xv, 6–7, 10–11, 116, 118, 154, 168, 182, 194, 252, 274–6

Reid, Roger, 175

reptile, 2–3, 22, 25, 27, 40, 47–8, 50–1, 54, 57, 60, 62, 73, 75–6, 93, 98, 108, 129, 135, 137–8, 142, 149, 169, 177, 184, 186, 188, 196, 198, 216, 269, 287–8, 293, 297–8

Relihan, Bruce, 112–3, 165

reptile-like amphibian, 4, 26, 31, 102, 184, 196–7, 199, 297

resting trace, xii, 22, 205–7, 225–30, 303, 313

Rindsberg, Andrew K., 73, 127, 143, 149, 155, 162, 164–5, 168, 179, 182, 274, 292

Robledo Mountains, 133–6, 139, 150, 156–7, 211, 215, 270, 310

Rusophycus, 225, 228–31, 280, 307

sauropod, 39–40, 47

scale tree, 46, 49, 72, 76, 131, 243–5, 247, 300

scorpion traces, 9, 42, 219, 230–2, 237, 251, 301

search image, 149

seed fern, 5, 21, 49, 68, 70–1, 128, 131, 239, 246–7, 302. *See also* pteridosperms

siderite, 44

Sigillaria: 28, 70, 113, 239; *elegans*, 242–3, 282; Devils Tower specimen, 28

Simpson, George Gaylord, xii, xvi, 2–3, 97–100, 273, 285, 291

SMR. *See* Surface Mining and Reclamation Division

sphenophyte, 239, 247, 298, 302

spider, viii, 9, 42, 169, 230, 232, 236, 280, 300

splitter, 96, 135–6. *See also* lumper

spoil pile, 30–1, 70, 95, 101–2, 114–7, 121, 156, 252

State Lands Division, vii, viii, 171, 173, 176, 249, 253, 276–7

Steven C. Minkin Paleozoic Footprint Site. *See* Minkin Paleozoic Footprint Site

Stiaria, 43, 112–3, 126–7, 207–15, 220–1, 264, 267, 280, 301–2

Stigmaria, 28, 46, 70

straddle, 41, 251

stratigraphic section, 117–20, 227

stride, 41, 185, 197, 251, 261, 302

succession. *See* stratigraphic section

sun orbit, 81–2

superposition, law of, 15, 115

Surface Mining and Reclamation Division, 171

Surface Mining Control and Reclamation Act, 153–4, 167–8, 170, 175